IFCoLog Journal of Logics and their Applications

Volume 4, Number 1

January 2017

Disclaimer

Statements of fact and opinion in the articles in IfCoLog Journal of Logics and their Applications are those of the respective authors and contributors and not of the IfCoLog Journal of Logics and their Applications or of College Publications. Neither College Publications nor the IfCoLog Journal of Logics and their Applications make any representation, express or implied, in respect of the accuracy of the material in this journal and cannot accept any legal responsibility or liability for any errors or omissions that may be made. The reader should make his/her own evaluation as to the appropriateness or otherwise of any experimental technique described.

ISBN 978-1-84890-237-4
ISSN (E) 2055-3714
ISSN (P) 2055-3706

College Publications
Scientific Director: Dov Gabbay
Managing Director: Jane Spurr

http://www.collegepublications.co.uk

Printed by Lightning Source, Milton Keynes, UK

SCOPE AND SUBMISSIONS

This journal considers submission in all areas of pure and applied logic, including:

pure logical systems
proof theory
constructive logic
categorical logic
modal and temporal logic
model theory
recursion theory
type theory
nominal theory
nonclassical logics
nonmonotonic logic
numerical and uncertainty reasoning
logic and AI
foundations of logic programming
belief revision
systems of knowledge and belief
logics and semantics of programming
specification and verification
agent theory
databases

dynamic logic
quantum logic
algebraic logic
logic and cognition
probabilistic logic
logic and networks
neuro-logical systems
complexity
argumentation theory
logic and computation
logic and language
logic engineering
knowledge-based systems
automated reasoning
knowledge representation
logic in hardware and VLSI
natural language
concurrent computation
planning

This journal will also consider papers on the application of logic in other subject areas: philosophy, cognitive science, physics etc. provided they have some formal content.

Submissions should be sent to Jane Spurr (jane.spurr@kcl.ac.uk) as a pdf file, preferably compiled in LaTeX using the IFCoLog class file.

Contents

ARTICLES

PREFACE

Previous editions of *Tools for teaching logic* (TTL) took place as follows. The first TTL conference took place in 2000[1]. It was the idea of an international group of logicians that, in 1998, created ARACNE, a European Community ALFA (America Latina Formación Académica) network. The second conference took place in 2006 (`http://logicae.usal.es/SICTTL/`), and the third in 2011 (`http://logicae.usal.es/TICTTL/`).

Proceedings of the 2000 edition were published in [3] and those of the 2006 were published in [4]. For the latter edition, a special volume of the *Logic Journal of the IGPL* was published in 2007 [5]. Finally, selected contributions to the third conference were published in *Lecture Notes in CS* [6]. As we write this, the number of downloads of the latter are over 16 thousand.

The *Fourth International Conference on Tools for Teaching Logic*, TTL 2015 (http://ttl2015.irisa.fr/) was held during June 9-12 2015 at *Institut de Recherche en Informatique et Systèmes Aléatoires* (IRISA), Rennes, France. This special issue contains nine papers that have been selected from the 35 papers of the conference [1]:

1. Anne Zamansky. Teaching Logic to Information Systems Students: a Student-centric Approach

2. Gilles Dowek. Rules and derivations in an elementary logic course

3. Johan van Benthem An old discipline with a new twist: the course "Logic in Action"

4. Jørgen Villadsen, Alexander Birch Jensen, Anders Schlichtkrull NaDeA: A Natural Deduction Assistant with a Formalization in Isabelle

5. María Manzano, Nitsa Movshovitz-Hadar, Diane Resek Leon Henkin: A logician's view on mathematics education

6. Ryo Hatano, Katsuhiko Sano and Satoshi Tojo Teaching Modal Logic from The Linear Algebraic Viewpoint

7. Patrick Blackburn. The New Trivium

---

[1] aracne.usal.es/congress/congress.html

8. John Slaney. Logic considered fun

9. Jeremy Seligman and Declan Thompson. Teaching natural deduction in the right order with Natural Deduction Planner

The contributions can be typed according to two main streams that are not exclusive:

- Teaching logic to a specific audience: 1, 2, 3, 5, 6, 7.

- Softwares for teaching logic: 4, 6, 8, 9.

We thank all the authors who have contributed to this special issue. Each submission for this special issue were assigned to two reviewers. We are immensely grateful to Giovanna D'Agostino, Carlos Areces, Éric Badouel, Sandrine Blazy, David Cachera, Susanna Epp, Annie Foret, Tim French, Thomas Genet, Andreas Herzig, Colin de la Higuera, Josje Lodder, Emiliano Lorini, Étienne Lozes, Valeria de Paiva, David Pichardie, Ram Ramanujam, Olivier Ridoux, Martin Strecker and Moshe Vardi for their valuable efforts and support.

We are also very grateful to the Editorial Board of IfCoLog for agreeing in publishing this volume. Thanks are due to Jane Spurr, M. Antonia Huertas Sánchez, Dov Gabbay and Michael Gabbay.

We would like to also thank again all the reviewers for the conference TTL 2015 as well as all the members of the organizing committee of TTL2015, M. Antonia Huertas Sánchez, Joao Marcos, María Manzano and Elisabeth Lebret, for their valuable help for this volume to exist.

We hope that this volume will inspire teachers so that the way logic is taught to a wide range of students from undergraduate to postgraduate gets improved, and thereby make those audiences aware of the tremendously importance of this topic at the crossing of many established disciplines.

September 2016,

Sophie Pinchinat & François Schwarzentruber

References

[1] Huertas A., Marcos J., Manzano M., Pinchinat S., Schwarzentruber F. (eds) Proceedings of the Fourth International Conference on Tools for Teaching Logic. Université de Rennes 1. 2015

[2] ASL Committee on Logic Education. Guidelines for Logic Education. Bull. Symb. Logic, 1(1):4-7, 1995. URL = http://www.ucalgary.ca/aslcle/guidelines

[3] Manzano, M. Proceedings of the First International Congress on Tools for Teaching Logic. University of Salamanca, 200. DL: S.443-2000. URL = http://logicae.usal.es

[4] Manzano, M. Pérez-Lancho, B. Gil, A. Proceedings of the Second International Congress on Tools for Teaching Logic. University of Salamanca, 2006. URL = http://logicae.usal.es

[5] Ditmarsch, H. and Manzano, M. (eds) Special Issue: Tools for Teaching Logic. Logic Journal of the IGPL. 15(4). 2007

[6] Blackburn, P. Ditmarsch, H. Manzano, M. and Soler-Toscano, E. (eds) Tools for Teaching Logic. Lecture Notes in Computer Science. 6680. 2011.

[7] Manzano, M., Sain, I. Alonso, E. (eds) The Life and Work of Leon Henkin: Essays on His Contributions. Birkauser/Springer. 2014

 Received 11 October 2016

Teaching Logic to Information Systems Students: a Student-centric Approach

Anna Zamansky[*]

Information Systems Department, University of Haifa, Israel
annazam@is.haifa.ac.il

Abstract

In contrast to Computer Science and Software Engineering, where it's fundamental role is widely recognized, logic plays a practically non-existent role in information systems (IS) curricula. In this paper we argue that instead of logic's exclusion from the IS curriculum, a significant adaptation of the contents, as well as teaching methodologies, is required for an alignment with the needs of IS practitioners. We present a vision for such adaptation and report on concrete steps towards its implementation in the design and teaching of a course for graduate IS students at the University of Haifa. We discuss the course plan and present a qualitative analysis of feedback provided by students of the course.

Keywords: LOGIC EDUCATION, COMPUTING CURRICULUM.

1 Introduction

Numerous works addressed the role of logic and formal methods in the computing disciplines curricula. While its importance as a core discipline for computer scientists and software engineers is widely acknowledged (see, e.g., the ACM curriculum guidelines [19, 13]), there is a growing frustration of educators concerning a poor alignment of such courses to the needs of practitioners of the computing disciplines. In [16], J. Makowsky writes: "Teachers of logic in Computer Science are often still teaching courses which are a mix of formalizing logical reasoning, meta-mathematics and the fading reverberations of the famous crisis of the foundations of mathematics. By doing so, they are contributing to the disappearance of their courses from mainstream undergraduate education in Computer Science. We have to rethink which aspects of logic matter for the Computer Science undergraduate programs."

[*]Supported by The Israel Science Foundation under grant agreement no. 817/15.

According to J. Wing [22]: "...we still face the educational challenge of teaching mathematical foundations like logic and discrete mathematics to practicing or aspiring software engineers. We need to go beyond giving the traditional courses and *think about who the target students are.*"

While a poor alignment of the logic courses to the needs of practitioners seems to be acknowledged (see also [3] in this context), works offering student-centric approaches, i.e., practical advice on *how* to make logic courses more effective and relevant for computing practitioners and/or reporting on empirical studies in logic education are scarce. A notable exception for the latter is the Beseme project ([17]): in a three-year study, empirical data on the achievements of two student populations was collected: those who studied discrete mathematics (including logic) through examples focused on reasoning about software, and those who studied the same subject illustrated with more traditional examples. An analysis of the data revealed significant differences in the programming effectiveness of these two populations in favor of the former.

In this paper[1] we take a *student-centric* approach in the sense of providing practical advice, supported by student feedback analysis, for a particular target audience, namely Information Systems students. The (academic) field of information systems (IS) encompasses two broad areas: (i) acquisition, deployment, and management of information technology resources and services, and (ii) development and evolution of infrastructure and systems for use in organization processes. Thus, as opposed to computer science (CS), IS's primary focus is on an organization's mission and objectives and the application of information technology to further these goals. Yet the IS discipline shares a significant body of knowledge with CS and software engineering (SE), reflected also in the intersection of the respective study programs' curricula. Logic, however, does not appear to be in this intersection – almost none of the IS undergraduate study programs include such course in their curriculum. As opposed to the ACM CS [19] and SE [13] curriculum guidelines, the analogous guidelines for IS [21] do not refer to logic as a core mathematical discipline.

The current state of affairs is suboptimal for several reasons. First of all, most of the reasons for including logic in the CS/SE curricula still hold in the IS domain. Secondly, the lack of experience with formal notation forms a major cognitive barrier to the adoption of formal methods by IS practitioners ([28]). This is further reinforced by the fact that because many IS study programs tend to be marketed as programs "excluding the hard math", the students come to see the lack of mathematical courses as a benefit, and express disappointment[2] when any formal notations

---

[1]This is a revised and extended version of [25] and [24].

[2]Quoting one of our graduate students who was assigned to read a research paper on formal methods: "When I see formal definitions, I just want to cry." Notably, she is one of the best students

are integrated in the IS core courses – thus creating a vicious circle.

Although a typical IS major may need a less extensive mathematical background than a CS major, it is our view that rather than excluding logic from the IS curriculum, a significant adaptation is needed to align it with IS objectives. In this paper we provide some practical suggestions for how to adapt logic courses to the context of IS by reporting on our experience in designing and teaching the course "Logic and Formal Specification" to graduate students at the Information Systems (IS) department at the University of Haifa, which includes a *mandatory* course on logic and formal methods in its graduate study program. We discuss our view of *what* should be included in the IS logic toolbox (analogously to the CS logic toolbox of [15]). Finally, we present the results of a pilot study on the students' own perceptions of the importance and benefits of the course. The study was carried out by administering an open-ended quesionnaire to 23 students in the years 2013-2014. A qualitative analysis of the collected data reveals that the course' benefits reported by students are mainly perceived as improvements in cognitive processes, such as analytical and abstract thinking, decomposition of problems and modelling.

2   The IS Logic Toolbox

The main practical objective in teaching logic to IS practitioners is providing them with the ability to apply formal methods in industry. Application of formal aspects is particularly important for software quality control, i.e., activities for checking (by proof, analysis or testing) that a software system meets specifications and that it fulfills its intended purpose.

Due to the density of the IS curricula, one currently cannot afford to have one course on pure formal logic and then another on formal methods. This problem is also discussed in [23] in the context of CS. Therefore, one must develop a mixture which combines introductory formal logic together with an introduction to the formal methods relevant for the IS domain. In what follows we briefly survey previous reflections on the content of logic and formal methods courses that practitioners *really* need and their integration into the curricula, and propose how to adapt those ideas for the context of IS.

2.1   Relevant Approaches

Recently there has been an ongoing discussion about whether the traditional logic syllabus for CS is relevant for practitioners. We start by briefly outlining some

---

in her class.

relevant proposals (mostly in the context of CS), the ideas of which are close in spirit to the vision we present below.

In [15, 16], J. Makowsky questions the suitability of the standard logic syllabus to the needs of CS practitioners. He states: "The current syllabus is often justified more by the traditional narrative than by the practitioner's needs." He further notes that most classical logic textbooks follow the narrative of the rise and fall of Hilbert's program, emphasizing the following ideas:

- Logic is needed to resolve the paradoxes of set theory;

- First-order logic (FOL) is the most important logic due to its completeness theorem;

- The main theorems of FOL are the completeness and compactness theorems;

- The tautologies of FOL are not recursive;

- One cannot prove consistency within rich enough systems.

This, according to Makowsky, is *not* what a CS practitioner needs: "The proof of the Completeness Theorem is a waste of time at the expense of teaching more the important skills of understanding the manipulation and meaning of formulas." What a practitioner needs is to:

- understand the meaning and implications of modeling the environment in terms of precise mathematical objects and relations;

- understand and be able to distinguish the intended properties of this modeling and its side-effects;

- be able to discern different level of abstraction, and

- understand what it means to prove properties of modeled objects.

In her papers [23, 22], J. Wing stresses the importance of integrating formal methods into the existing CS curriculum by teaching their common conceptual elements, including state machines, invariants, abstraction, composition, induction, specification and verification. She gives discrete mathematics and mathematical logic as crucial prerequisites.

The above proposals on *what* to teach are extremely relevant for IS practitioners. On the question of *how* to teach, the paper "Integrating Formal Methods into Computer Science Curricula at a University of Applied Science" ([20]) of Tavolato and Vogt offers some useful insights. It discusses teaching formal methods at universities of applied sciences, where there are usually limiting factors which are relevant to the IS context as well: (i) students have very limited theoretical background, and (ii) they are strongly focused on the direct applicability of what they are taught. In this context the authors stress the importance of making the practical applicability of the theory understandable to students, and making use of real industry-inspired examples.

In what follows, we extend and adapt the above proposals for the context of IS, and provide our vision on aligning the teaching of logic to the needs of IS practitioners.

2.2 Making Logic Relevant for IS

2.2.1 The What

Logic is a prerequisite for understanding and successfully using formal methods, which in their turn can significantly contribute to software quality control. We agree with the view taken in [23] that the main basic formal conceptual elements with which the students need familiarizing include state machines, abstraction, composition, induction, invariants, specification and verification. While the students encounter the concepts of state machines, abstraction and composition in other IS courses (such as modeling and design), aspects related to working with formal specifications are not covered elsewhere in the curriculum. However, an IS practitioner needs to:

- read, write and understand formal specifications;

- formalize informal specifications;

- analyze specifications and detect sources of incompleteness, inconsistency and complexity;

- reason about specifications, and

- check a system against a specification.

Thus, by adapting and extending the previous proposals of CS logic education to the context of IS, we arrive at the following IS logic toolbox:

1. Basic principles for reasoning about sets;

2. Use of induction and invariants;

3. Propositional and first-order logic and their axiomatizations;

4. Formal specification and verification techniques and methodologies.

2.2.2 The How

As to *how* to teach logic to IS students, i.e., designing concrete teaching methodologies, the following considerations need to be taken into account:

- *Creating links to software domain.*
  Although it has been believed for some time that studying logic improves software development skills, this common belief has recently been empirically validated by a study in [17]. As pointed out by [20], software related examples are also useful for increasing the motivation of students, who can see the applications of the studied material in the domain of their interest. In [27] a particular example of establishing such software-related link is described, in which teaching sequent calculi was supported via a hands-on assignment in software testing.

- *Integrating education methodologies and tools.*
  The integration of methodologies and tools from education could be beneficiary in a number of aspects. First of all, empirical studies show that the use of formal methods poses objective difficulties for practitioners ([4, 8]). They are also hypothesized to be a major hindering factor for the acceptance of formal methods in industry ([28]). The difficulties students experience when studying logic and formal methods ([20]) could be addressed using studies of cognition, which could provide insights into students' mental processes when studying formal concepts. For instance, several studies analyzed gaps between students' intuition and formal thinking in mathematics (see, e.g., [6]) and students' difficulties in handling abstraction ([9, 10]).

- *Hiding some of the complexity.* Exposing the students to the full intricate complexities of mathematical logic (such as a full proof of the completeness theorem, or dealing with variables not free for substitution) has the potential to confuse novices struggling to understand new ideas. However, most IS practitioners will not encounter these complexities in industry. This is in line with the research agenda of indirect application of formal methods ([12]), which calls for hiding intricate complexities behind automatic tools with intuitive user interface. Similar benefits for programming are also mentioned in [17].

3 Teaching Logic for IS

In this section we demonstrate how the vision presented above has been implemented in the design of our course "Logic and Formal Specification". The course has been taught at the IS department at the University of Haifa for several years by the author[3]. The course is a mandatory course for graduate students, and its length is one semester, 4 hours per week.

3.1 Course Description

Below we provide a short description of the course's main topics, together with textbooks from which the material is adapted. The course is divided into two parts. The first part covers the basics of formal logic, and includes the following topics:

Part I: Introduction to Logic

- *Informal laws of mathematical reasoning*

Exercise: Show that for all sets A, B, C: if $A \subseteq B$ and $B \subseteq C$, then $A \subseteq C$.

Law 1 - universal statements: If you want to prove a statement about all things of a certain kind, choose an arbitrary thing and show that the statement holds for it.

Let A, B, C be arbitrary sets.

Law 2 - conditional statements:
If you want to prove a statement of the form "if x then y", assume x and use it to prove y.

Suppose that $A \subseteq B$ and $B \subseteq C$. We now prove $A \subseteq C$...

Figure 1: (Informal) Laws of Reasoning and a Demonstration of their Application

Our starting point is the place where the students left off in discrete mathematics course: basic set-theoretical concepts. However, our primary focus is not on understanding the concepts themselves, but on *reasoning* about them by applying informal logical laws. Accordingly, the students are asked to provide proofs of basic claims, explaining which laws were used at each stage. A

---

[3]Perhaps it is important to mention here the author's relevant background. She is an associate professor at the Information Systems Department at the University of Haifa with active research interests in applied logic and more than 10 years of experience in teaching logic and formal methods to various student audiences.

Example question: n processes run a computer program which performs the operation x^{++}, *where x is a variable shared by all processes. The ++ operation is mapped into three sequential sub-operations:*

1. *load x to register*

2. *increment the value in the register*

3. *store the register value to x*

The programmer intended to write a program, such that at the end of its execution the value of x is n. What other scenarios are possible, given that the processes can run in parallel?

Figure 2: An Example of Inductive Reasoning for Proving a Program Property

basic example is provided in Figure 1; the students are asked to justify every step of their proof.

The presentation of the informal laws and other proof tips is adapted from David Makinson's textbook "Sets, Logic and Math for Computing" [14], which is also one of the official textbooks of the course. The informal laws become explicit at the object level when classical propositional and first-order logic are introduced to the students (e.g., the law for proving general statements can be captured by the rule inferring $\forall x \psi$ from $\psi(x)$, and the law for proving conditional statements is captured by the deduction theorem.) At this stage we revisit the proofs and pinpoint the application of these laws.

- *Induction:* mathematical, structural and computational induction.
  Structural induction is at the heart of a number of formal concepts relevant for verification and validation of software: fixed point constructions, model checking, program analysis and many more. Therefore a special emphasis is put on the topic throughout the course. Starting with a motivational example for an inductive set (by presenting the MU puzzle by Douglas Hofstadter ([11])), we provide formal definitions of induction and exemplify their use by proving properties of programs. One example is provided in Figure 2: one can prove, e.g., that if at the beginning of execution $x = 0$, then at the end of execution x may assume at least one of the values $0, 1, ..., n$. Other examples presented in class are adapted from Chapter 2 of *Foundations of Computer Science* of Aho and Ullman ([2]).

- *Classical Propositional and First-Order Logic:* syntax and semantics, satisfiability and validity, Hilbert-style axiomatization, formalization of natural lan-

12

guage sentences.

For this part of the course we mostly adapt parts of the standard presentation of most mathematical logic textbooks. We place a special emphasis on the inductive definitions of the set of wffs of CPL and the set of theorems of the Hilbert-style axiomatization, and show several examples of proofs using induction (including the deduction theorem). This comes at the expense of omitting the proofs of the completeness and compactness theorems (in line with the recommendation of [15]).

- *Survey of non-classical logics:* temporal logic, modal logic, many-valued logic, fuzzy logic, non-monotonic logic, paraconsistent logic.

  This part of the course is implemented by requiring each of the students to deliver a short presentation on a non-classical logic of his choice. While the importance of temporal logic in this context is perhaps the most obvious one due to its well-known applications in verification, other non-classical logics also have IS-relevant applications (see, e.g., [7, 5, 29]). The goal is to increase the awareness of the students to the immense variety of logics outside the realm of classical logics.

Part II: Introduction to Formal Specification

This part of the course builds up on the knowledge obtained at the previous part. The final aim is for the students to be able to understand and write formal specifications using the Z notation. For this we have adapted the material from the textbook [18], covering the basic aspects of Z: types, schemas and reasoning about Z specifications.

4   Students Feedback Analysis

The course has only been taught in its current form for three years. While making decisive conclusions about its effectiveness is perhaps premature, an important dimension in evaluating such effectiveness is the students' acceptance and reaction to it. To gain a better understanding of these factors, a preliminary qualitative study was undertaken by administering a questionnaire, which was filled by twenty-three students who took the course in the years 2013-2014.

Recall that the limiting factors typical of the target audience are in many respects similar to those described in [20]. The first is *lack of mathematical background*: the undergraduate IS study program at the University of Haifa does not include a course in logic, and the majority of students have only a background in discrete mathematics, where they are taught very basic concepts of set theory. The second limiting factor is their *lack of motivation*: the majority of the students return to graduate school several years after receiving their B.A, while working full-time. They typically expect the topics to be directly relevant to their IS practice, and usually exhibit difficulty in coping with the dense and abstract material taught in the course. In light of these factors, we were expecting some of the students to claim, basically, that the course was too hard without being helpful for their future as IS practitioners. However, only one student out of 23 felt the course was not useful for his practice.

In what follows we describe the results of an exploratory study exploring perceptions of twenty three students over the years 2013-2014. This sample included 8 female and 15 male students; 12 students out of 22 had no prior experience in industry. The questionnaire included the following open-ended questions.

Q1 Is it important for practitioners whose work is related to software development to study logic and formal methods? Why?

Q2 In what way (if at all) is the course's content useful for <u>information systems</u> practitioners?

Q3 What (if at all) were the course's contributions for you personally?

Q4 How relevant was the background from your Discrete Mathematics course? In what way (if at all) was it helpful?

Q5 In what ways would you recommend that we improve the course?

In what follows we focus mainly on the answers received to questions Q2 and Q3. Only three students responded that logic and formal methods are not useful (Q2):

1. I worked at two different places in industry, and never have I seen the courses' content put to any use...

2. It is not necessary for software development.

3. It depends on the work environment. I think it's not useful.

Two of them also thought the course was not useful for them personally (Q3).

Out of those who responded positively to both questions, one of the most striking observations was the extensive use of formulations related to mental processes, such as "thinking", in particular "analytical/logical thinking" in answers to both questions.

E.g., answers to question Q2 included:

1. It improves **thinking** about problem modeling.

2. I think that it opens directions for **thinking** about how things really work under the surface.

3. The world of software is based on understanding the needs and modeling them in precise terms. Many such models require **logical thinking**.

4. The course's contents develop and deepen **ways of thinking**.

5. The course helps shaping **thinking** that can help in programming.

6. The course improves **analytical thinking**.

7. The course is very helpful in improving **thinking that is not necessarily algorithmic**. A different one, out of the box.

8. Of course! **Correct and systematic thinking** of IS practitioners helps in requirements specification.

Notably, no participants provided concrete examples of direct use of the courses' content in answering Q2. Yet several of them took a confident stand when speaking of their own personal experience in Q3:

1. I have already applied the new skills at work, using truth tables and proofs.

2. It improved my modeling skills. I'm certain!

3. I am now using the tools when reading scientific papers.

4. I was surprised to see how helpful the tools we studied are in practice.

Moreover, when answering question Q3, several participants referred again (implicitly or explicitly) to an improvement in their *mental* processes:

1. The course introduced order into complex topics. It gave me tools to simplify complex problems and find easy and efficient solutions.

2. It made me think in a modular way, providing me with the ability to grasp more complex models.

3. It improved my ability to refer to problems schematically.

4. It provided me with an abstract view on the problems of software design.

	Q2 (general IS practitioner)	Q3 (personal experience)
thinking	8	8
understanding	7	8
formulation	5	3
modelling	1	0
research	0	3
general knowledge	0	5

Table 1: Categories emerging from answers to Q2 and Q3 and number of students using each category

5. It made me realize there are systematic solutions to problems that seem unsolvable at first.

6. I learned to reduce complex problems to simpler ones.

Table 1 summarizes the main skill categories that emerged during text analysis of questions Q2 and Q3, providing the number of students that used formulations related to these categories. We intend to use these categories as a basis for further, deeper quantitative investigation that will hopefully provide evidence for the benefits of teaching logic courses to IS future practitioners.

5 Summary and Future Research

While there recently has been quite a lot of discourse on the poor alignment of logic courses to the practical needs of computing practitioners, practical "field" advice on how the situation can be improved is still very scarce. The current paper makes a contribution in this direction in the context of the target population of IS students, for whom the lack of direct relevance of the traditional logic courses seems to have led to their (unfortunate) exclusion from the undergraduate curriculum. And yet logic remains central to IS objectives, as it is the key to applying formal methods in specification, verification and validation of information systems. Therefore, further empirical evidence in the spirit of the Beseme project ([1]) is needed to convince decision makers that such courses are useful for IS practitioners. To be successful this will involve taking more *student-centric* approaches, which involve understanding the impact of teaching logic on students' achievements, as well as their perceptions and attitudes. Moreover, overcoming the objective difficulties of students with logic and formal methods could be made easier by integrating new technologies for enhancing education. One step in this direction was taken in [26], where the potential of using online social networks such as Facebook for teaching logic is explored.

Based on our experience teaching the Logic and Formal Specification course to graduate IS students, we feel that using software-related and comprehensible examples, and simplifying logical intricacies contributes to achieving the courses' objectives. In addition, student feedback showed positive perceptions of the benefits of taking the course, which are mainly related to general cognitive processes (as opposed to specific skills and/or tools). Categories which have emerged from a qualitative analysis of this feedback can be adapted for new and more detailed survey instruments which we hope will provide decision makers with (much needed) evidence for the benefits of the inclusion of logic in the IS undergraduate curriculum.

Another planned future research project is an empirical investigation of *how* to make formal specification more understandable for students. This question is particularly interesting due to its direct relation to the more general topic of comprehensibility of specifications. In this context we plan to develop a tool for automatic analysis of Z specifications, which will then be used for empirical evaluations.

We hope that this paper will start a wider discussion on *what* logical background is needed for IS practitioners and *how* it should be taught. We further hope that this will lead to a logic textbook with an IS-orientation, which would be a welcome addition to the large existing variety of CS-oriented books.

References

[1] Beseme website, http://www.cs.ou.edu/ beseme/.

[2] Alfred V Aho and Jeffrey D Ullman. *Foundations of computer science*, volume 2. Computer Science Press New York, 1992.

[3] Douglas Baldwin, Henry M Walker, and Peter B Henderson. The roles of mathematics in computer science. *ACM Inroads*, 4(4):74–80, 2013.

[4] Deirdre Carew, Chris Exton, and Jim Buckley. An empirical investigation of the comprehensibility of requirements specifications. In *Empirical Software Engineering, 2005. 2005 International Symposium on*, pages 10–pp. IEEE, 2005.

[5] Hendrik Decker. A case for paraconsistent logic as foundation of future information systems. In *CAiSE Workshops (2)*, pages 451–461. Citeseer, 2005.

[6] Lisser Rye Ejersbo, Uri Leron, and Abraham Arcavi. Bridging intuitive and analytical thinking: Four looks at the 2-glass puzzle. *For the Learning of Mathematics*, 2014.

[7] Neil A Ernst, Alexander Borgida, John Mylopoulos, and Ivan J Jureta. Agile requirements evolution via paraconsistent reasoning. In *Advanced Information Systems Engineering*, pages 382–397. Springer, 2012.

[8] Kate Finney. Mathematical notation in formal specification: Too difficult for the masses? *Software Engineering, IEEE Transactions on*, 22(2):158–159, 1996.

[9] Orit Hazzan. How students attempt to reduce abstraction in the learning of mathematics and in the learning of computer science. *Computer Science Education*, 13(2):95–122, 2003.

[10] Orit Hazzan and Rina Zazkis. Reducing abstraction: The case of school mathematics. *Educational Studies in Mathematics*, 58(1):101–119, 2005.

[11] Douglas R Hofstadter. Godel, Escher, Bach. *New Society*, 1980.

[12] Heinrich Hussmann. Indirect use of formal methods in software engineering. In *ICSE-17 Workshop on Formal Methods Application in Software Engineering Practice, Seattle (WA), USA. Proceedings*, pages 126–133. Citeseer, 1995.

[13] Richard J LeBlanc, Ann Sobel, Jorge L Diaz-Herrera, Thomas B Hilburn, et al. *Software Engineering 2004: Curriculum Guidelines for Undergraduate Degree Programs in Software Engineering*. IEEE Computer Society, 2006.

[14] David Makinson. *Sets, logic and maths for computing*. Springer, 2012.

[15] Johann A Makowsky. From Hilbert's program to a logic toolbox. *Annals of Mathematics and Artificial Intelligence*, 53(1-4):225–250, 2008.

[16] Johann A Makowsky. Teaching logic for computer science: Are we teaching the wrong narrative? In *Fourth International Conference on Tools for Teaching Logic, TTL 2015 Proceedings*, 2015.

[17] Rex L Page. Software is discrete mathematics. In *ACM SIGPLAN Notices*, volume 38, pages 79–86. ACM, 2003.

[18] Ben Potter, David Till, and Jane Sinclair. *An introduction to formal specification and Z*. Prentice Hall PTR, 1996.

[19] Mehran Sahami, Mark Guzdial, Andrew McGettrick, and Steve Roach. Setting the stage for computing curricula 2013: computer science–report from the acm/ieee-cs joint task force. In *Proceedings of the 42nd ACM technical symposium on Computer science education*, pages 161–162. ACM, 2011.

[20] Paul Tavolato and Friedrich Vogt. Integrating formal methods into computer science curricula at a university of applied sciences. In *TLA+ Workshop at the 18th International Symposium on Formal Methods, Paris, Frankreich.*, 2012.

[21] Heikki Topi, Joseph S Valacich, Ryan T Wright, Kate Kaiser, Jay F Nunamaker Jr, Janice C Sipior, and Gert-Jan de Vreede. Is 2010: Curriculum guidelines for undergraduate degree programs in information systems. *Communications of the Association for Information Systems*, 26(1):18, 2010.

[22] Jeannette M. Wing. Teaching mathematics to software engineers. In *Algebraic Methodology and Software Technology, 4th International Conference, AMAST '95, Montreal, Canada, July 3-7, 1995, Proceedings*, pages 18–40, 1995.

[23] Jeannette M Wing. Invited talk: Weaving formal methods into the undergraduate computer science curriculum. In *Algebraic Methodology and Software Technology*, pages 2–7. Springer, 2000.

[24] Anna Zamansky and Eitan Farchi. Helping the tester get it right: Towards supporting agile combinatorial test design. In *2nd Human-Oriented Formal Methods workshop (HOFM 2015)*, 2015.

[25] Anna Zamansky and Eitan Farchi. Teaching Logic to Information Systems Students: Challenges and opportunities. In *Fourth International Conference on Tools for Teaching Logic, TTL 2015 Proceedings*, 2015.

[26] Anna Zamansky, Kiril Rogachevsky, Meira Levy, and Michal Kogan. How many likes can we get for logic? Exploring the potential of Facebook for enhancing core software engineering courses. In *Proceedings of the European Conference on Software Engineering Education*, 2016.

[27] Anna Zamansky and Yoni Zohar. Mathematical does not mean boring: Integrating software assignments to enhance learning of logico-mathematical concepts. In *International Conference on Advanced Information Systems Engineering*, pages 103–108. Springer, 2016.

[28] Marc K Zimmerman, Kristina Lundqvist, and Nancy Leveson. Investigating the readability of state-based formal requirements specification languages. In *Proceedings of the 24th International Conference on Software engineering*, pages 33–43. ACM, 2002.

[29] Didar Zowghi, Vincenzo Gervasi, and Andrew McRae. Using default reasoning to discover inconsistencies in natural language requirements. In *Software Engineering Conference, 2001. APSEC 2001. Eighth Asia-Pacific*, pages 133–140. IEEE, 2001.

Received 11 October 2016

Rules and derivations in an elementary logic course

Gilles Dowek
Inria and ENS Cachan
61, avenue du Président Wilson 94235 Cachan Cedex, France.
gilles.dowek@ens-cachan.fr

When teaching an elementary logic course to students, who have a general scientific background, but have never been exposed to logic, we have to face the problem that the notions of deduction rule and of derivation are completely new to them, and are related to nothing they already know, unlike, for instance, the notion of model, that can be seen as a generalization of the notion of algebraic structure, or the notion of computable function, that is a particular case of the notion of function.

We present, in this paper, a strategy to introduce these notions: start with the notion of inductive definition [1], then, the notion of derivation comes naturally. We also show, with three examples: computability theory, automata theory, and proof theory, that derivations are pervasive in logic—we could have given more examples in formal grammars, rewriting, etc. Thus, defining precisely this notion at an early stage is a good investment to later define other notions. Finally, we show that we need to distinguish two notions of derivation: that of *derivation labeled with elements* and that of *derivation labeled with rule names*.

In this paper, no proofs are given, and not even precise definitions. These can be found, for instance, in [2].

1 From inductive definitions to derivations

1.1 A method to define sets: inductive definitions

Inductive definitions are a way to define subsets of a set A. The inductive definition of a subset P is formed with a family of functions f_1, from A^{n_1} to A, f_2, from A^{n_2} to A, etc. These functions are called *rules*. For example, the function $f_1 = \langle \rangle \mapsto 0$, from $\mathbb{N}^0$ to $\mathbb{N}$, and $f_2 = \langle x \rangle \mapsto x + 2$, from $\mathbb{N}^1$ to $\mathbb{N}$ are rules.

Instead of writing these rules $f_1 = \langle \rangle \mapsto 0$ and $f_2 = \langle x \rangle \mapsto x + 2$, we often write them

$$\overline{0}\ f_1$$

$$\frac{x}{x+2} \, f_2$$

But despite this new notation, rules are things the students already know: functions.

These rules define a function F from $\mathcal{P}(A)$ to $\mathcal{P}(A)$

$$F(X) = \bigcup_i \{f_i(a_1, ..., a_{n_i}) \mid a_1, ..., a_{n_i} \in X\}$$

For example, the two rules above define the function

$$F(X) = \{0\} \cup \{x + 2 \mid x \in X\}$$

and, for instance, $F(\{4, 5, 6\}) = \{0, 6, 7, 8\}$, $F(\varnothing) = \{0\}$, $F(\{0\}) = \{0, 2\}$, etc.

The function F is monotonic and continuous. Thus, it has a smallest fixed point P, which is the inductively defined subset of A. This smallest fixed point can be defined in two ways

$$P = \bigcap_{F(X) \subseteq X} X = \bigcup_i F^i(\varnothing)$$

The first definition characterizes the set P as the smallest set closed by f_1, f_2, etc. the second as the set containing all the elements that can be built with these functions in a finite number of steps.

The notion of monotonicity and continuity of a function from $\mathcal{P}(A)$ to $\mathcal{P}(A)$ can then be introduced and the two fixed point theorems can be proved with mathematically oriented students. They can be admitted otherwise.

Continuing with our example, the set P of even numbers can be characterized as the smallest set containing 0 and closed by the function $x \mapsto x + 2$, or as the union of the sets $\varnothing$, $F(\varnothing) = \{0\}$, $F^2(\varnothing) = \{0, 2\}$, $F^3(\varnothing) = \{0, 2, 4\}$, etc.

1.2 Derivations

A *derivation* is a tree whose nodes are labeled with elements of A and such that if a node is labeled with b and its children with a_1, ..., a_n, then there exists a rule f such that $b = f(a_1, ..., a_n)$. A *derivation of an element a* is a derivation whose root is labeled with a. We can then prove, by induction on i, that all the elements of $F^i(\varnothing)$ have a derivation. The property is trivial for $i = 0$. If it holds for i and $b \in F^{i+1}(\varnothing)$, then by definition $b = f(a_1, ..., a_n)$ for some rule f and $a_1 \in F^i(\varnothing)$, ..., $a_n \in F^i(\varnothing)$, thus, by induction hypothesis, a_1, ..., a_n have derivations. Hence, so does b.

Thus, from the second property $P = \cup_i F^i(\varnothing)$, we get that all elements of P have derivations. Conversely, all elements that have a derivation are elements of P.

Continuing with our example the number 4 has the derivation

$$\dfrac{\overline{\dfrac{\overline{0}}{2}}}{4}$$

1.3   Rule names

There are several alternative definitions of the notion of derivation. For instance, when $b = f(a_1, ..., a_n)$, instead of labelling the node just with b, we can label it with the ordered pair formed with the element b and the name of the rule f. For instance, the derivation of 4 above would then be the tree

$$\dfrac{\overline{\dfrac{\overline{\langle 0, f_1 \rangle}}{\langle 2, f_2 \rangle}}}{\langle 4, f_2 \rangle}$$

more often written

$$\dfrac{\overline{\dfrac{\overline{0}\,f_1}{2}\,f_2}}{4}\,f_2$$

Such a derivation is easier to check, as checking the node

$$\dfrac{2}{4}$$

requires to find the rule f such that $f(2) = 4$, while checking the node

$$\dfrac{2}{4}\,f_2$$

just requires to apply the rule f_2 to 2 and check that the result is 4.

But these rules names are redundant, as soon as the relation $\cup_i f_i$ is decidable. So, in general, they can be omitted.

23

1.4 Derivations labeled with rule names

Instead of omitting the rule names, it is possible to omit the elements of A. The derivation of 4 is then the tree

$$\frac{\overline{f_1}}{\dfrac{f_2}{f_2}}$$

that can also be written

$$\frac{\quad}{\quad} f_1$$
$$\frac{\cdot}{\quad} f_2$$
$$\frac{\cdot}{\quad} f_2$$
$$\cdot$$

We introduce this way a second kind of derivations *labeled with rules names*. In contrast, the previous derivations can be called *labeled with objects*.

Although it is not explicit in the derivation, the element 4 can be inferred from this derivation with a top-down *conclusion inference algorithm*, because the rules f_i are functions. The conclusion of the rule f_1 can only be $f_1(\langle\rangle) = 0$, that of the first rule f_2 can only be $f_2(\langle 0 \rangle) = 2$, and that of the second can only be $f_2(\langle 2 \rangle) = 4$.

More generally, for each derivation labeled with rule names D, there is at most one a such that D is a derivation of a, the existence of such an a can be decided and, when it exists, this a can be computed from D. As a consequence, the set of ordered pairs $D : a$ such that D is a derivation of a is decidable.

1.5 Proof-terms and type systems

Derivations labeled with rules names are often written as a term, that is in a linear form. For instance the derivation

$$\frac{\quad}{\quad} f_1$$
$$\frac{\cdot}{\quad} f_2$$
$$\frac{\cdot}{\quad} f_2$$
$$\cdot$$

is often written $f_2(f_2(f_1))$. Such a term is called a *proof-term*.

The decidable set of ordered pairs $\pi : a$ such that π is a proof-term of a can itself be defined by an inference system obtained from the original one by replacing each rule

$$\frac{a_1 \ \dots \ a_n}{b} \, R$$

24

with the rule

$$\frac{\pi_1 : a_1 \, ... \, \pi_n : a_n}{R(\pi_1, ..., \pi_n) : b} \, R$$

In our example, we get the rules

$$\frac{}{f_1 : 0} \, f_1$$

$$\frac{\pi : a}{f_2(\pi) : a + 2} \, f_2$$

and the ordered pair $f_2(f_2(f_1)) : 4$ has the derivation

$$\frac{\dfrac{\overline{f_1 : 0}}{f_2(f_1) : 2}}{f_2(f_2(f_1)) : 4}$$

This second inference system is called a *type system*. It defines a decidable set—it is even an automaton in the sense of [4].

Moreover the conclusion inference algorithm transforms into a *type inference algorithm*. For each proof-term π, there is at most one a such that $\pi : a$ is derivable, the existence of such an a can be decided and, when it exists, this a can be computed from π.

1.6 Making the rules functional

Natural deduction proofs [6, 5], for instance, are often labeled both with sequents and rule names, for instance

$$\frac{\overline{P, Q, R \vdash P} \, \text{axiom} \quad \overline{P, Q, R \vdash Q} \, \text{axiom}}{P, Q, R \vdash P \wedge Q} \, \wedge\text{-intro}$$

but they can be labeled with sequents only

$$\frac{\overline{P, Q, R \vdash P} \quad \overline{P, Q, R \vdash Q}}{P, Q, R \vdash P \wedge Q}$$

and proof-checking is still decidable. They can also be labeled with rule names only, but we have to make sure that all the deduction rules are functional, which is often not the case in the usual presentations of Natural deduction. The rule

$$\frac{\Gamma \vdash A \quad \Gamma \vdash B}{\Gamma \vdash A \wedge B} \, \wedge\text{-intro}$$

25

is functional: there is only one possible conclusion for each sequence of premises, but the axiom rule

$$\overline{\Gamma, A \vdash A} \text{ axiom}$$

is not. To make it functional, we must introduce a different rule axiom$_{\langle \Gamma, A \rangle}$ for each ordered pair $\langle \Gamma, A \rangle$. Thus, the proof above must be written

$$\cfrac{\overline{P, Q, R \vdash P} \text{ axiom}_{\langle \{Q,R\}, P \rangle} \qquad \overline{P, Q, R \vdash Q} \text{ axiom}_{\langle \{P,R\}, Q \rangle}}{P, Q, R \vdash P \wedge Q} \wedge\text{-intro}$$

And from the derivation labeled with rule names

$$\cfrac{- \text{ axiom}_{\langle \{Q,R\}, P \rangle} \qquad - \text{ axiom}_{\langle \{P,R\}, Q \rangle}}{\cdot} \wedge\text{-intro}$$

the conclusion $P, Q, R \vdash P \wedge Q$ can be inferred.

In a linear form, this derivation is $\wedge$-intro(axiom$_{\langle \{Q,R\}, P \rangle}$, axiom$_{\langle \{P,R\}, Q \rangle}$) and its type, $P, Q, R \vdash P \wedge Q$, can be inferred.

2 Derivations in elementary computability theory

2.1 A pedagogical problem

The set of computable functions is often defined inductively, as the smallest set containing the projections, the null functions, and the successor function, and closed by composition, definition by induction, and minimization.

But to study the computability of properties of computable functions, we need a secondary notion of *program*, that is we need a way to express each computable function with a expression of a finite language, to which a Gödel number can be assigned. A usual solution is to introduce Turing machines, λ-calculus, or any other language at this point.

This solution however is not pedagogically satisfying as, while the students are still struggling to understand the inductive definition of the set of computable functions, we introduce another, that is often based on completely different ideas, letting them think that logic made of odds and ends. Moreover, the equivalence of the two definitions requires a tedious proof.

Such a second definition is in fact not needed as the inductive definition itself already gives a notion of program, through the notion of derivation.

2.2 Programs already exist

The function $x \mapsto x + 2$ is computable because it is the composition of the successor function with itself. But the derivation labeled with objects

$$\frac{\overline{x \mapsto x + 1} \qquad \overline{x \mapsto x + 1}}{x \mapsto x + 2}$$

cannot be used as a program, because to label its nodes, we would need a language to express all the functions, and there is, of course, no such language.

But if we use a derivation labeled with rule names instead

$$\frac{\overline{} \; Succ \qquad \overline{} \; Succ}{} \; \circ_1^1$$

and write the derivations in a linear form: $\circ_1^1(Succ, Succ)$, we obtain a simple variable-free functional programming language, to express the programs. We can introduce this way a symbol π_i^n for the n-ary i-th projection, $\circ_p^n(f, g_1, ...g_p)$ for the composition of the n-ary functions $g_1, ..., g_p$ with the p-ary function f, and $\mu^n(f)$ for the minimization of the $n + 1$-ary function f over its last argument, etc.

For instance, introducing a Gödel numbering $\ulcorner . \urcorner$ for these programs, and assuming there is an always defined function h such that

- $h(p, q) = 1$ if $p = \ulcorner f \urcorner$ and f defined at q

- and $h(p, q) = 0$ otherwise,

we get a contradiction: the function

$$k = \circ_1^1(\mu^1(\pi_1^2), \circ_2^1(h, \pi_1^1, \pi_1^1))$$

is defined at $\ulcorner k \urcorner$ if and only if it is not.

We get this way a proof of the undecidability of the halting problem that requires nothing else than the inductive definition of the set of computable functions.

3 Derivations in elementary automata theory

When introducing the notion of finite automaton, we often introduce new notions, such as those of transition rules and recognizability. Having introduced the notion of derivation from the very beginning of the course permits to avoid introducing these as new notions.

Consider for instance the finite state automaton

$$odd \xrightarrow{a} even \qquad even \xrightarrow{a} odd$$

where the state *even* is final. In this automaton, the word *aaa* is recognized in *odd*. Indeed

$$odd \xrightarrow{a} even \xrightarrow{a} odd \xrightarrow{a} even$$

If, instead of introducing a new notion of transition rule, we just define transition rules as deduction rules

$$\frac{even}{odd} \, a \qquad \frac{odd}{even} \, a \qquad \frac{}{even} \, \varepsilon$$

then, the element *odd* has a derivation

$$\cfrac{\cfrac{\cfrac{\cfrac{}{even} \, \varepsilon}{odd} \, a}{even} \, a}{odd} \, a$$

If we label this derivation with rule names we obtain

$$\cfrac{\cfrac{\cfrac{\cfrac{}{-} \, \varepsilon}{-} \, a}{-} \, a}{-} \, a$$

which can be written in linear form $a(a(a(\varepsilon)))$, or *aaa*. Thus, a word w is recognized in a state s if and only if it is a derivation, labeled with rule names, of s.

Transforming this inference system into a type system, like in Section 1.5, we get

$$\frac{w : even}{aw : odd} \, a \qquad \frac{w : odd}{aw : even} \, a \qquad \frac{}{\varepsilon : even} \, \varepsilon$$

And a word w is recognized in a state s if and only $w : s$ is derivable.

This example introduces a point that needs to be discussed: the rules

$$\frac{even}{odd} \, a \qquad \frac{odd}{even} \, a$$

are labeled with the same name. If the automaton is deterministic, we can replace these two rules with one: a function such that $a(even) = odd$ and $a(odd) = even$.

But for non deterministic automata, we either need to extend the notion of rule name, allowing different rules to have the same name, or to consider two rule names

$$\frac{even}{odd}\, a_1 \qquad \frac{odd}{even}\, a_2 \qquad \frac{}{even}\, \varepsilon$$

and map the derivation $a_1(a_2(a_1(\varepsilon)))$ to the word $a(a(a(\varepsilon)))$ with the function $|.|$ defined by: $|\varepsilon| = \varepsilon$, $|a_1(t)| = a(|t|)$, and $|a_2(t)| = a(|t|)$.

4 Introducing the Brouwer-Heyting-Kolmogorov correspondence

4.1 A radical change in viewpoint?

The Brouwer-Heyting-Kolmogorov interpretation, and its counterpart, the Curry-de Buijn-Howard correspondence, are often presented as a radical change in viewpoint: proofs are not seen as trees anymore, but as algorithms.

But, of course, these algorithms must be expressed in some language, often the lambda-calculus. Thus, proofs are not really algorithms, but terms expressing algorithms, and such terms are nothing else than trees. So, it is fairer to say that, in the Brouwer-Heyting-Kolmogorov interpretation, proofs are not derivation trees, but trees of a different kind. For instance, the tree

$$\frac{\dfrac{\overline{P \wedge Q \vdash P \wedge Q}}{P \wedge Q \vdash Q} \qquad \dfrac{\overline{P \wedge Q \vdash P \wedge Q}}{P \wedge Q \vdash P}}{\dfrac{P \wedge Q \vdash Q \wedge P}{\vdash (P \wedge Q) \Rightarrow (Q \wedge P)}}$$

is replaced by the tree

$$\frac{\dfrac{\overline{x}}{snd} \qquad \dfrac{\overline{x}}{fst}}{\dfrac{\langle,\rangle}{\lambda x : P \wedge Q}}$$

often written in linear form: $\lambda x : P \wedge Q\ \langle snd(x), fst(x) \rangle$.

4.2 Derivation trees labeled with rule names

Instead of using this idea of expressing proofs as algorithms, let us just try to label the derivation above with rule names. Five rules are used in this proof. Three of them are functional

$$\frac{\Gamma \vdash A \quad \Gamma \vdash B}{\Gamma \vdash A \wedge B} \wedge\text{-intro}$$

$$\frac{\Gamma \vdash A \wedge B}{\Gamma \vdash A} \wedge\text{-elim1}$$

$$\frac{\Gamma \vdash A \wedge B}{\Gamma \vdash B} \wedge\text{-elim2}$$

Let us just give them shorter names: $\langle,\rangle$, fst, and snd. The rule

$$\frac{\Gamma, A \vdash B}{\Gamma \vdash A \Rightarrow B} \Rightarrow\text{-intro}$$

is functional, as soon as we know which proposition A in the left-hand side of the antecedent is used. So, we need to supply this proposition A in the rule name, let us call this rule λA. Finally, the rule

$$\frac{}{\Gamma, A \vdash A} \text{axiom}$$

is functional, as soon as we know Γ and A. We could supply Γ and A in the rule name. However, we shall just supply the proposition A and infer the context Γ. Let us call this rule $[A]$. So, the proof above can be written

$$\frac{\dfrac{\overline{P \wedge Q \vdash P \wedge Q}\,[P \wedge Q]}{P \wedge Q \vdash Q}\,snd \quad \dfrac{\overline{P \wedge Q \vdash P \wedge Q}\,[P \wedge Q]}{P \wedge Q \vdash P}\,fst}{\dfrac{P \wedge Q \vdash Q \wedge P}{\vdash (P \wedge Q) \Rightarrow (Q \wedge P)}\,\lambda P \wedge Q}\,\langle,\rangle$$

and if we keep rule names only

$$\frac{\dfrac{-\,[P \wedge Q]}{\,\dot{-}\,}\,snd \quad \dfrac{-\,[P \wedge Q]}{\,\dot{-}\,}\,fst}{\dfrac{\dot{}}{\,\dot{-}\,\lambda P \wedge Q}}\,\langle,\rangle$$

which, in linear form is the proof-term $\lambda P \wedge Q \ \langle snd([P \wedge Q]), fst([P \wedge Q]\rangle\rangle$.

Transforming this inference system into a type system, like in Section 1.5, we get

$$\frac{\Gamma \vdash \pi : A \quad \Gamma \vdash \pi' : B}{\Gamma \vdash \langle \pi, \pi' \rangle : A \wedge B} \wedge\text{-intro}$$

$$\frac{\Gamma \vdash \pi : A \wedge B}{\Gamma \vdash fst(\pi) : A} \wedge\text{-elim1}$$

$$\frac{\Gamma \vdash \pi : A \wedge B}{\Gamma \vdash snd(\pi) : B} \wedge\text{-elim2}$$

$$\frac{\Gamma, A \vdash \pi : B}{\Gamma \vdash \lambda A \ \pi : A \Rightarrow B} \Rightarrow\text{-intro}$$

$$\frac{}{\Gamma, A \vdash [A] : A} \text{ axiom}$$

in which the ordered pair $\lambda P \wedge Q \ \langle snd([P \wedge Q]), fst([P \wedge Q]) \rangle : (P \wedge Q) \Rightarrow (Q \wedge P)$ is derivable. This is the scheme representation [3] of this proof.

Let us show that the conclusion can be inferred, although we have not supplied the context Γ in the axiom rule. The conclusion inference goes in two steps. First we infer the context bottom-up, using the fact that the conclusion has an empty context, and that all rules preserve the context, except λA that extends it with the proposition A

$$\frac{\dfrac{\overline{P \wedge Q \vdash .} \ [P \wedge Q]}{P \wedge Q \vdash .} snd \quad \dfrac{\overline{P \wedge Q \vdash .} \ [P \wedge Q]}{P \wedge Q \vdash .} fst}{\dfrac{P \wedge Q \vdash .}{\vdash .} \lambda P \wedge Q} \langle , \rangle$$

Then, the right-hand part of the sequent can be inferred with a usual conclusion top-down inference algorithm, using the fact that the rules are functional

$$\frac{\dfrac{\overline{P \wedge Q \vdash P \wedge Q} \ [P \wedge Q]}{P \wedge Q \vdash Q} snd \quad \dfrac{\overline{P \wedge Q \vdash P \wedge Q} \ [P \wedge Q]}{P \wedge Q \vdash P} fst}{\dfrac{P \wedge Q \vdash Q \wedge P}{\vdash (P \wedge Q) \Rightarrow (Q \wedge P)} \lambda P \wedge Q} \langle , \rangle$$

4.3 Brouwer-Heyting-Kolmogorov interpretation

In the rule

$$\frac{\Gamma, A \vdash B}{\Gamma \vdash A \Rightarrow B} \Rightarrow\text{-intro}$$

instead of supplying just the proposition A, we can supply the proposition A and a name x for it. Then, in the axiom rule

$$\frac{}{\Gamma, A \vdash A} \text{ axiom}$$

31

instead of supplying the proposition A, we can just supply the name that has been introduced lower in the tree for it. We obtain this way the tree

$$
\frac{\dfrac{-x}{\vdots} \; snd \qquad\qquad \dfrac{-x}{\vdots} \; fst}{\vdots \; \langle , \rangle}{\vdots \; \lambda x : P \wedge Q}
$$

in linear form $\lambda x : P \wedge Q \; \langle snd(x), fst(x) \rangle$, which is exactly the representation of the proof according to the Brouwer-Heyting-Kolmogorov interpretation.

So, the Brouwer-Heyting-Kolmogorov interpretation boils down to use of derivations labeled with rule names plus two minor modifications: context inference and the use of variables. These two modifications can be explained by the fact that Natural deduction does not really deal with sequents and contexts: rather with propositions, but, following an idea initiated in [7], some rules such as the introduction rule of the implication dynamically add new rules, named with variables.

References

[1] P. Aczel, An introduction to inductive definitions, *Handbook of Mathematical Logic*, Studies in Logic and the Foundations of Mathematics 90, 1977, pp. 739-201.

[2] G. Dowek, *Proofs and Algorithms: An Introduction to Logic and Computability*, Springer-Verlag, 2011.

[3] G. Dowek and Y. Jiang, On the expressive power of schemes, *Information and Computation*, 209, 2011, pp. 1231-1245.

[4] G. Dowek and Y. Jiang, Decidability, Introduction Rules and Automata, *Logic for Programming, Artificial Intelligence, and Reasoning*, 2015, pp. 97-111. Long version submitted to publication.

[5] J.-Y. Girard, Y. Lafont and P. Taylor, *Proofs and Types*, Cambridge University Press, 1990.

[6] D. Prawitz, *Natural deduction: A proof-theoretical study*, Almqvist & Wicksell, 1965.

[7] P. Schroeder-Heister, A natural extension of natural deduction, *The Journal of Symbolic Logic*, 49, 4, 1984, pp. 1284-1300.

Received 11 October 2016

An old discipline with a new twist: the course "Logic in Action"

Johan van Benthem

University of Amsterdam, Stanford University and Tsinghua University

Abstract

What are the basic logical notions and skills that all beginning students should learn, and that might stay with them as a useful cultural travel kit for their lives, even when an overwhelming majority will not become professional logicians? The course "Logic in Action" http://www.logicinaction.org/ tries to convey the idea that logic is about reasoning but also much more: including information and action, both by individuals and in multi-agent settings, studied by semantic and syntactic tools, and still confirming to the standards of precision of an exact and mathematized discipline. Viewed in this way, modern logic sits at a crossroads of academic disciplines where interesting new developments occur every day. In this light introduction, I explain the main ideas behind the design of the course, which combines predicate logic with various modal logics, and I lightly discuss its current manifestations and dialects in Amsterdam, Beijing and the Bay Area, as well as its future as an EdX pilot course.

1 History of the course

There is a thriving international market of new on-line logic courses today, witness the many projects presented at the successive TTL conferences [1] and the links there to earlier conferences in this series. Roughly speaking these endeavors fall into two kinds. Sometimes the new technology is used to create high-tech versions of largely standard fare in the traditional curriculum with, say, sophisticated graphics interfaces for classical natural deduction proof systems, like a Latin Mass with rock

---

I thank the organizers of the Conference on Tools for Teaching Logic, Rennes 2015, for giving me an opportunity and a forum for reflecting on the course "Logic in Action". I also thank the members of the core LiA development team for the course as well as the users that we know of, and finally, I am grateful to the two referees for this paper for providing very useful critical comments.

[1]See the website http://ttl2015.irisa.fr of these conferences.

guitars.[2] But sometimes also, there is ideological fervor behind the effort: the course designers have a special research agenda with their own view of logic, modifying or changing existing curricula, and they want to export their revolution by by-passing the academic colleagues and instead of that, influencing the youth.[3]

The course Logic in Action falls in the second activist category, and we will put our cards on the table in a moment. The course arose in the education group of the Spinoza Award project "Logic in Action" (1997–2002; http://www.illc.uva.nl/lia/) of the Dutch Science Organization, and it received a crucial further push by a grant from the Dutch Ministry of Economic Affairs in its program Creative Technologies meant to improve the national information infrastructure.

2 The general idea: a broader scope for logic

Traditional logic courses emphasize the study of correct inference patterns as the core business of logic, with propositional and predicate logic as paradigms of the methodology for doing so. Students are trained in basic skills which typically include translating natural language sentences into formulas, performing validity tests such as truth tables and tableaux, and often also, calculi for formal deduction.

Some problems with traditional courses   In our view, this traditional agenda is not neutral: it instills a large number of attitudes, often as hidden presuppositions. Let us identify a few of its subliminal messages.

First, inference is made the central concern of logic – but this move seems quite debatable. Inference or proof is just one topic in logic, and just as important are two other main themes: definability and computation, a point made already in the seminal Beth 1963 reflecting on the history of logic as well as its modern branches of proof theory, model theory and recursion theory.

Next, there is little reflection on what intellectual assets are actually activated by training in formula translation or formal proof. It is unclear whether there is any transfer to broader reasoning skills, and it may be significant that research logicians themselves never seem to use them in their meta-theory. Criticisms of this didactic kind have in fact occurred throughout the last century: a modern study of transfer

---

[2]This is how I would view, e.g., the popular and very well-designed course "Logic & Proofs" at Carnegie Mellon University, http://oli.cmu.edu/courses/free-open/logic-proofs-course-details/.

[3]This activist stance is what I see in the Stanford course "Language, Proof and Logic" (http://online.stanford.edu/course/language-proof-and-logic) inspired by situation theory, and in the more logic-programming and resolution-based open-domain CS course "Introduction to Logic" (https://www.coursera.org/learn/logic-introduction). But their designers may feel very differently!

of skills also involving experimental cognitive studies is Haskell 2000.[4]

Next, the usual emphasis on formal proof somehow suggests that mathematical activities are the highest point of logical intellectual skills, a claim as debatable as thinking that the best test of someone's moral fiber is her behavior in church. Reasoning in down-to-earth practice, with its open universes of relevant considerations, tells us much more about what logical rationality a person can bring to bear.

Finally, the standard emphasis on teaching complete logical systems as the locus of logic is a very peculiar methodology, different even from the problem solving skills taught in mathematics and science courses. One comes for a logical formula or two in the store (just as we learn a few crucial and generally helpful algebraic equations), hoping that it will help us through some crucial steps in a problem-solving argument. But instead, one finds that one has to buy a system, a huge infinite supply of valid patterns, and worry about their staying fresh for years.

Broadening the scope   Raising the preceding concerns does not mean that there is something inherently wrong with the traditional curriculum in logic, as far as it goes – only that the discipline of logic has much wider scope than what this standard agenda of topics might suggest. The major aim of the course 'Logic in Action' is conveying this broader picture from the start as being much more true to what logic is today and what its range is across the university and elsewhere. If we do not get this across at base level, students will either not see what logic is really good for, or, they will develop a narrow conception of the field which then keeps them locked afterwards into biased philosophical or mathematical conceptions.

Logic as information handling   One way of achieving this mind-opening is by shifting the emphasis from inference alone to the study of a much broader range of informational activities as the subject of logic. Besides inference, such logical activities also include making observations and doing experiments, asking questions and processing answers to them, and engaging in communication generally. Therefore, the course 'Logic in Action' treats two realms on a par, purely deductive inference, and intelligent conversation, as highlighted to our students in the following picture of Euclid's "Elements" versus Rubens' painting 'The Philosophers':

---

[4]An emphasis on isolated formal activities need not be harmless, it may even make enemies. I have often observed this in interdisciplinary circles where colleagues from other fields who went through a logic course became firmly convinced of the Scholasticism and irrelevance of our discipline.

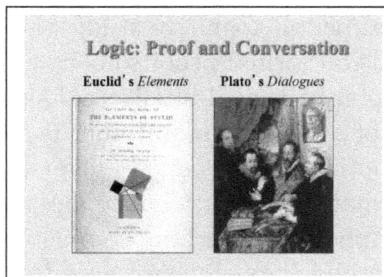

Interestingly, our themes are already present in ancient Chinese logic, witness a key dictum in the Moist School (500–300 BC; cf. Graham 2003) that knowledge comes from three sources: hearing from others, demonstration, and experience.

Histories   As it happens, while not neglecting the essential Greek origins, this quotation is highlighted in our course with a side purpose: also make it clear to students that by learning logic, they become part of a worldwide cultural stream, not just ancestor worship of Greek Antiquity. The course has many such historical sidebars, all aiming at installing some more general erudition.

But we also emphasize that inference and observation are information sources on a par in modern science, where we need the two in tandem to understand our world. And in more playful mode, to the classical lonely thinker with eyes closed and ears shut, we juxtapose the detective Sherlock Holmes whose success shows that, far from the usual view of logic as organized pedantry, logical skills are not just duties that we perform, but also talents that we appreciate and that even give us pleasure.

Two strands: structure of the world, structure of human activity   Even so, the course presents no criticism of other views. One general way of thinking about what is said here goes back to a pervasive feature of logic throughout its history, and something that even surfaces in many hardcore textbooks. One can think of logic as describing the most general structure of reality and its inventory of atomic, negative, or disjunctive facts, individual and general facts. In that metaphysical sense, logic would be there even if there were no human beings at all, as on the cold and lonely planets we see in astronomical documentaries. One can soften this perspective a bit in terms of objective information available about and in the world (another view of logic that can be found in prominent textbooks, cf. Devlin 1991), but again this information would be there even if there were no human agents picking it up.

But there is also another stream, right from the ancient Greek origins of the discipline with Aristotle and Plato (but also prominent in the Chinese tradition), of logic as manifesting itself in activities of conversation, dialogue and debate, whether cooperative or competitive. On this agency view, logical laws are about moves and strategies that agents have toward winning in dialogue games, and the very logical constants now correspond with structured actions in argumentation or conversation. On this second view, then, communication and strategic interaction are crucial to logic, and the patterns described by logical systems may just as well be forms of rational behavior as forms of language as patterns forming the grooves of our world. 'Logic in Action' emphasizes the second view as much as the first. [5]

An interdisciplinary cross-roads   This view comes with a broad canvas of disciplines that modern logic interacts with. While students in many disciplinary courses taught today, be they mathematicians, philosophers, or linguists, may be told that logic is typically 'theirs' (with only rumors of lapses into other fields), the reality of the field today is that it interacts with, feeds into, and is inspired by contacts with the old interfaces of philosophy and mathematics, but just as much with computer science, linguistics, and in recent years also some cognitive science. Probably most logic research today takes place in computer science, including some of the most innovative frontiers. Thus, in this course, computation in a broad sense is highlighted as a core concern of logic, and a running theme next to proof or definability.

---

[5]Of course, the two views are not in conflict. In the end, structured activity that does not fit the structure of the world may not have much of a chance from an evolutionary perspective.

37

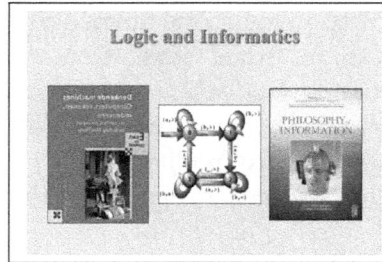

This is the intellectual environment that we convey to students in this course. Logic is one's ticket to broadmindedness, not to one particular disciplinary lifestyle.

3   And so: teaching a broader range of logical skills

In terms of paradigmatic logical acts, then, the basic repertoire to be taught gets extended. Say, a question is as basic a logical act as an inference. And likewise, an interactive strategy is as important as a proof, say, as a way of guiding communication or argumentation. But how do we make all this concrete to students?[6]

New standard example: Three Cards   An appealing aspect of this multi-agent interactive view is that set pieces of logical reasoning to be taught now become much more interesting and appealing to students (and adults) than the usual simple syllogisms about Socrates' mortality or Boolean inferences about which box the keys are in. Here is a typical challenge, somewhat of a classic by now. Much of our basic reasoning in daily life is not just about the facts, but it also crucially involves what we know about what others do or do not know. Here is a scenario that was once played out with real children in the Amsterdam science museum "NEMO".

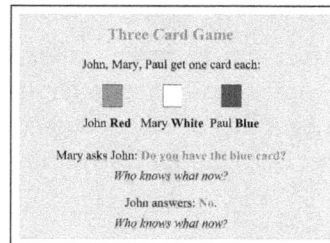

"The Cards". Three cards 'red', 'white', 'blue' are given to three children: 1, 2, 3, one to each. This fact is common knowledge in the whole group.

---

[6]A stream of research on 'dynamic-epistemic logic' forms the backdrop to this line in the course: cf. van Ditmarsch, van der Hoek & Kooi 2007, van Benthem 2011, and van Benthem 2014.

The children see their own cards, not those of the others. The actual distribution over 1, 2, 3 is 'red, white, blue' (written **rwb**. Now a conversation takes place. Child 2 asks 1: "Do you have the blue card?" Then 1 answers truthfully: "No". Who knows what during this conversation?

This scenario always generates classroom interaction, including mistaken claims. Here is the logical answer. Assuming that questions and answers are sincere (not unrealistic with children), 2 indicates that she does not know the answer, and so she cannot have the blue card. This tells 1 at once what the deal was. But 3 did not learn, since he already knew that 2 does not have blue. When 1 says she does not have blue, this now tells 2 the deal. 3 still does not know the deal; but since he can perform the reasoning just given, he knows that the others know it.

Humans often go through this sort of reasoning, with different knowledge for different agents acting as the driving force for communication. Indeed, puzzles like this pose challenges that people worldwide find interesting, witness the discussion of the solution of the 'Cheryl Birthday Puzzle',[7] a knowledge problem that went viral in the spring of 2015 after having appeared on a talk-show in Singapore.

Cheryl's birthday is one of 10 possible dates.

May 15	May 16	May 19
June 17	June 18	
July 14	July 16	
August 14	August 15	August 17

Cheryl tells the month to Albert and the day to Bernard.

Albert says. "I don't know the birthday, but I know Bernard doesn't know either."
Bernard then says. "I didn't know at first, but now I do know."
Albert then says. "Now I also know Cheryl's birthday."
When is Cheryl's birthday?

**When Is Cheryl's Birthday?
Answer To Viral Math Puzzle**

A new feature: modeling skills This is logic in action at a challenging level, including inferences, questions and answers. And it involves a further important skill not usually taught in introductions to logic, namely, the ability to model a given scenario in a concrete semantic manner. Indeed, it is not hard to make students see that we can model the initial situation for the Three Cards as a set of six alternatives (the possible deals of the cards), related by easily drawable labeled uncertainty lines for players, as in the leftmost diagram of the following sequence:

---

[7]See https://en.wikipedia.org/wiki/Cheryl27s_Birthday for details.

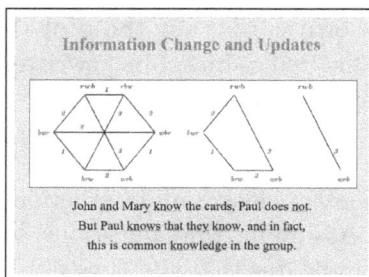

Information Change and Updates

John and Mary know the cards, Paul does not.
But Paul knows that they know, and in fact,
this is common knowledge in the group.

With this direct visual structure, child 2 cannot tell **rwb** and **bwr** apart if she finds herself in either of these deals, but when in those same situations, child 1 and 3 can. The information flow in the preceding example can then be made very concrete in terms of updates, leading to the next two diagrams in the picture:

> 2's question removes **rbw**, **wbr**, reducing the range to four options. Then 1's answer removes **bwr**, **brw**, and we are left with a final diagram **rwb**, **wrb**, in which it is directly visible that 1 and 2 know the cards, 3 does not, though 3 does know that the others know. The last fact is non-trivial information in itself, of a more social interactive nature.[8]

Extended desiderata   What skills and insights do we expect students to learn in the wider world of this course? Certainly, we do not want to give up on classical topics, since propositional and predicate logic with their standard agenda are still at the core of the field. Also, there is of course nothing wrong with the traditional virtues of logic education that come with this, such as increasing precision, appreciating the architecture of logical systems, and acquiring a sense of the beauty of abstract mathematical formulations. Indeed, such learning experiences also have to be, and can be, supplied for the further tasks mentioned here. This includes an understanding of the systematic theory behind the examples we have given.

But in addition, we want new topics that reflect the wider world of informational activities that we sketched, dealing with the logic of information, update, and interaction. And didactically, this set of topics requires modeling skills beyond the usual core. For instance, we do not want routine 'translation' of the natural language text of the Three Cards scenario into formulas, the way we drill students in a standard course to become little text processors. Such translations mix details of syntax with essentials for the task at hand – and true logical ability consists, more creatively, in

---

[8]Teaching unusual material like this challenges students in new ways. Recently, one observed that Child 1 does not even need to answer the question, but only has to say that he now knows the cards, and then Child 2 will know the cards as well. This then raised interesting general discussion in class about how epistemic information can replace factual information in communication.

picking out the important features only. In brief, we want a paraphrase into essential formulas, the way we also use just a handful of mathematical equations to model physics problems. And preferably, we want a semantic model or diagram for the setting, and based on that, an understanding of the relevant information flow.

And finally, in an information society like ours, the world of human reasoning is entangled ever more with computing technology, whose origins go back to logic in other historical channels. Accordingly, in terms of preparing them for life, we want the students to understand some basics of the computational structure and complexity of the informational processes that form the topic of this course.

Next, we say a bit more about the course resulting from all these desiderata.

4   Contents and chapter structure

How can we teach the above enlarged set of themes and skills? Perhaps the most obvious approach is to merely extend today's standard curriculum. In a way, our course has that feature. 'Logic in Action' has the following two main parts, with a third as a supplement for a more ambitious version.

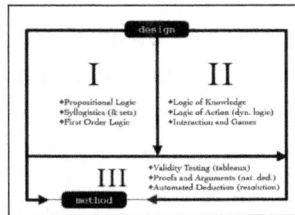

Part I   The first part of the course contains the basics of propositional logic, then the syllogistic as a first, historically but also systematically important, extension toward reasoning about objects and predicates, and finally, full first-order logic with quantifiers. These systems are presented as progressively richer ways of describing the world, be it physical space or conceptual space. The way we do this has a few new twists (see the description of our recurrent chapter structure given below), but we largely follow the standard agenda of basic topics. This part covers the descriptive dimension of logic that we mentioned in Section 2. Next, we turn to the activity dimension that we saw as a complementary view of what logic is about.

Part II   The second part of the course is then devoted to the main ingredients of information-driven agency. A first chapter on epistemic logic focuses on semantic modeling of information, including knowledge that agents have about facts and about knowledge of other agents. Next to get at the dynamics of the actions involved in communication, and agency in general, a chapter on dynamic logic of programs, and structured action generally, gives a standard base logic of structured computation. Finally, the two strands of information and action are brought together in a chapter on logic and games, as a grand finale where preferences come into the picture, as well as the fundamental notions of strategy and equilibrium.

Available materials   This brief survey article is not the place to give precise details of these six chapters, for which we refer the reader to the public-domain website

http://www.logicinaction.org/

and the free textbook "Logic in Action" and further download materials there:

http://www.logicinaction.org/docs/lia.pdf.

Part III   There is also a third optional part in the course, with technical material on major methods for proof and computation: semantic tableaux, natural deduction, and resolution. This is meant for students or teachers who have time to spare, or who just cannot let go of traditional themes. We also envisage adding a chapter on basic meta-theory offering acquaintance with proving important facts about logical systems, both classical and epistemic-dynamic. This material is more traditional again, emphasizing once more that we have no quarrel with standard curricula, and that the new systems in Part II still fall under a standard methodology.

Extension implies pruning   In all this, a choice had to be made in setting ambitions. If we keep the usual content of the first standard chapters the same, then a course like this will become top-heavy, and also, we miss an opportunity to remove historical clutter from the old curriculum. But if we rethink things more radically, then hard and perhaps controversial choices must be made. Do we still teach translation from natural language to formulas, with the usual drill? Do we teach formal deduction in detail, despite legitimate concerns about its broader transfer value to reasoning skills, or its adequacy as a model for what mathematical proof really is? If we stick to the standard course size, something has got to give. [9]

In our course, we have economized mainly on extensive translation drill, and on proof skills, though they are not gone completely. In particular, we have kept some axiomatic calculi to at least familiarize the students with the important intellectual idea of a symbolic uninterpreted systems view of deduction. Moreover, precisely

---

[9]This pruning may also have a positive value. People often forget that dropping worn-out topics from a curriculum can yield as much progress in a field as adding new ones.

because axiomatic proofs may involve surprising twists and shortcuts via lemmas undreamed of in more placid proof search environments like tableaux or sequent calculi, these pose more creative challenges to students. We have also economized on the usual formal set-theoretic presentation of model-theoretic semantics for first-order logic, which is often a stumbling block for students anyway, and which can also be questioned on technical logical grounds (Andréka, van Benthem, Bezhanishvili & Németi 2014). This set-theoretic garb also has the disadvantage of making first-order truth, which students already understand intuitively, look weird and exotic.

Note that the course does not become easier in this way, since the content structure for these topics in Part I now carries over to the new topics of Part II.

Coherence and chapter design Our broader agenda does have a didactic down side. The expanded set of topics runs a risk of incoherence and incongruity since its scope is so wide. Hence, to increase a sense of uniformity for the student, all chapters, no matter how different their topics, have been set up in a similar manner:

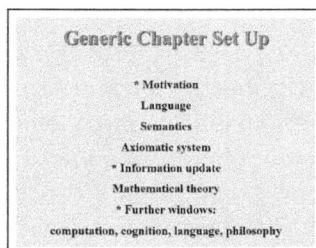

> **Generic Chapter Set Up**
>
> * Motivation
> Language
> Semantics
> Axiomatic system
> * Information update
> Mathematical theory
> * Further windows:
> computation, cognition, language, philosophy

As illustrated in the displayed picture, each chapter repeats the same pattern of sections – called Motivation, Language, Semantics, Axiomatic System, Update, Mathematical Theory, and Further Windows, where the latter are illustrations oriented toward the broader intellectual environment of logic in computation, cognition, language, and philosophy. Let us describe the generic structure of a few chapters, with less or more standard topics, to illustrate how this set-up works.

Chapter 1 In the chapter on propositional logic, 'motivations' are classifying of structure in the world as well as finding patterns in argumentation, 'language' introduces the idea of abstract symbolic syntax as a major historical achievement, and 'models' are of course the evergreen of truth tables. For an 'axiomatic system' we teach some Hilbert-style formula manipulation, which also gets students used to idea that finding proofs is not trivial. A new feature is teaching 'update' where new information decreases a current range of options, and where we show how some puzzles can be solved naturally either by deduction or by update to one single remaining option. This shows the semantics at work in a way that students find appealing, while the harmony of semantics and proof theory also features concretely. Next, in a section on 'mathematical theory' we introduce definability of connectives, as well as the notions of soundness and completeness for logical systems. Finally, 'further windows' in this case are toward the usual logic puzzles, but beyond that, mainly toward computation: networks for Boolean algebra, and complexity, including the $P = NP$ problem. After all, propositional logic is deeply connected with the emergence of computer science. Of course, the chosen illustrations in such windows can, and will touch on different disciplines in other chapters.

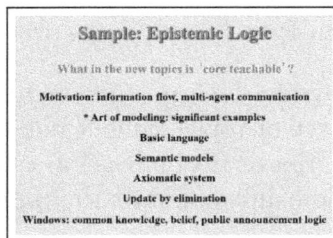

Sample: Epistemic Logic

What in the new topics is 'core teachable'?

Motivation: information flow, multi-agent communication

* Art of modeling: significant examples

Basic language

Semantic models

Axiomatic system

Update by elimination

Windows: common knowledge, belief, public announcement logic

Chapter 4 Now the very same structure is also used, say, in the chapter on epistemic logic. We motivate the issues by means of simple informational scenarios concerning questions and answers that students immediately find appealing.[10] Introducing a language with knowledge operators allows them to state significant things about the agents involved in such scenarios in a concise manner, and finding models for this language that match a given intuitive scenario then turns out to be an attractive non-trivial task. Thus, instead of routine drill, we now emphasize the 'Art of Modeling'. Axiomatic systems such as modal S5 now stand for significant (if often controversial, and always discussion-generating in class) properties of knowledge, and making concrete deductions shows surprising connections. Update is the way of solving puzzles like the Three Cards, discussed earlier, in a satisfying systematic manner. Mathematical theory includes again completeness, or, more ambitiously qua abstract ideas, an introduction to the notion of bisimulation as an invariance between information or process models. In terms of further outlooks, epistemic logic is well-suited to discussing basic themes in philosophy (say, adding belief, and then discussing the surplus of real knowledge over belief) and cognitive science, where interactive social 'Theory of Mind' is considered a typical human skill.

---

[10]In particular – referring to the first picture displayed here – normally, my asking you in a Beijing street whether the structure depicted is the ceremonial Central Gate of Tsinghua University tells you several important epistemic social things: (a) I do not know the answer, (b) I want to know the answer, and (c) I think that you may know the answer.

Chapters 5 and 6 In a similar manner, we structure the next chapter on dynamic logic as a stream-lined abstract modal version of the basic Hoare Calculus of structured programs and actions, and at the same time, as a natural companion to the epistemic logic chapter for the purpose of describing information dynamics. The perspectives of Chapters 4 and 5 then come together once more in the chapter on games, along two lines. We introduce logic games for earlier tasks of formula evaluation or proof, and we define game logics as revealing basic structures in reasoning about, and inside, social interaction. We also show some mathematical background such as Zermelo's Theorem and broader connections with game theory.

Windows and the range of logic As for the intended interdisciplinary range discussed in Section 2: our windows at the end of these chapters include topics such as computational content of logics (say, satisfiability checking as computation), information and the internet, natural language (for instance, generalized quantifiers are a window after the chapter on the syllogistic), cognitive science (the Wason Card Task and difficulties in actual reasoning, confronting logical systems with 'natural logic' in cognitive architecture), and some history of logic in other cultures, especially in China. Finally, we keep emphasizing the value of the mathematical aspects of logic, none of which are meant to be endangered by this course: precision in formulation, abstraction, systematicity, and the beauty of meta-theory. [11]

Having concluded our description of content and structure, let us now look at some issues of didactic implementation and concrete practical experiences with the course 'Logic in Action'. Does the above really work?

---

[11]Of course we cannot cover every technical aspect. For instance, most modal logics of Part II can be systematically translated into (decidable fragments of) the first-order logic of Part I. But even though, technically, this generates further coherence to the course, we feel that this translation theme would probably only confuse students at this early stage of their logical education.

5 The spirit and the letter of the course

The spirit There are various ways of looking at this course. Our educational goal is to teach the students modern ideas and skills they may find helpful in further academic professions, or even beyond in society. We try to strike many chords in doing so. At a basic level, we try to show that logic is fun, using classic puzzles as well as newer items such as Sudoku or the Cheryl Birthday Puzzle. It is just a simple fact that many people enjoy exercising logical thinking skills, and students are no exception. Next, we try to teach the students what we genuinely think are the core topics of the field: deduction, computation, information, and interaction, using a broadened set of topics that we hope will become standard. To us, an example like the Cards puzzle is as genuinely logical as worrying about Socrates' mortality. In addition, we try to convey an appealing picture of logic as a broad and lively evolving field that connects between many disciplines, or put more negatively: we try to combat widespread narrow exclusive views of logic by opening interdisciplinary and cross-cultural windows. In doing so, we also try to convey that logic still has a great future ahead of it, given that so much has kept happening over the last century. Finally, perhaps more silently, we also hope to convey a less utilitarian idea to the students: that logic has a cultural value in itself that enriches them.

The letter These are the high-sounding ideals. In subsequent sections we will discuss what happens when these meet educational practice. But right here, let us also state another perspective: if you wish, 'the letter' of this course, that seems to be what remains on many working colleagues' radar when they use the material presented here. Take away the above ideals, and just look at what has to be taught, the bottom-line of all courses in academic reality. One way of describing our curriculum is simply this: we add modal logic to the traditional topics of propositional and predicate logic. The rationale for this terse description is that modal logic is indeed the technical core underlying our added chapters on epistemic and dynamic logic. While this mathematical formulation is an outrageously one-dimensional projection of what is contained in the course, and a misleading one in several ways, it does have the virtue of being short and intelligible. Moreover, since the connections of modal logic to classical systems are well-understood, the addition fits very nicely, so hard-bitten illusion-free teachers can just see this as their task.

6 The internet dimension

As stated at the start, the original impetus for making this course happen (and a major motivation for its funding) was an initiative toward creating free courses available on the internet, and supported by new technology. Where do we stand?

On what there is Our material is freely available on our public website mentioned above: http://www.logicinaction.org/, in the form of a textbook, slides supporting class sessions, videos, and exercises from various sources with worked-out answers.

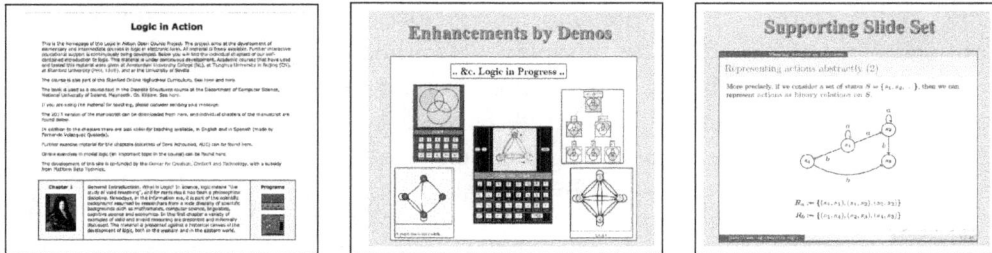

Overreach? Still, the ambitions in our team were much higher when the project started. We wanted to create a complete e-book with live links to background material, applets for specific tasks or demonstrations, and clickable windows for stepping right into the field of logic, from interfaces with automated deduction systems to more theoretical sources. A few chapters of this sort are indeed available on the above website, drawing on the innovative material developed by Jan Jaspars, a pioneer in computer-supported logic teaching in The Netherlands – for more samples, see, e.g., this website of the Dutch Open University:

https://www.ou.nl/web/logica-in-actie

Ideally, this electronic paradise would allow for complete self-study of 'Logic in Action' by worldwide users of the course, helped along by equally automated self-tests after chapters, without any interaction with human designers or teachers, except perhaps in the form of filmed lectures or video clips. In 2014, a more modest pilot version of this course, taught at Stanford University, was indeed formatted for the EdX platform in a preliminary way.[12]

None of this broader internet agenda has materialized seriously so far. We will discuss in Section 8 why this is so, and how much of a bad or good thing this is.

---

[12]See the material at http://explorecourses.stanford.edu/search;jsessionid=1uudpox4o4m9v13p6 nh1froldi?q=PHIL+150E%3A+Logic+in+Action%3A+A+New+Introduction+to+Logic&view= catalog&filter-coursestatus-Active=on&academicYear=20132014.

7 Experiences so far

This course has been taught in since 2008 in various classroom versions, mostly at an undergraduate level, at universities in Amsterdam, Beijing, Berkeley, Maynooth, Seville, and Stanford. Moreover, shorter versions of 'Logic in Action' have been taught occasionally elsewhere for more diverse audiences, including the 2011 ESSLLI Summer School in Ljubljana. An adapted version is also a standard part of the curriculum of the Stanford On-Line High School.

This fits the intended range as originally envisaged for this course material: from advanced high school level onward to lower university levels.

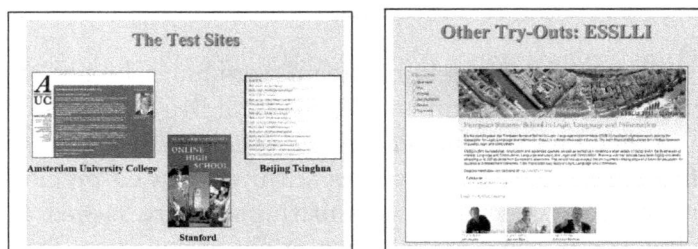

Except for the usual short-term student evaluations, there has been no systematic pedagogical evaluation yet of 'Logic in Action', which would have to involve a longer-term study of the intended lasting effects qua skills and attitudes.

Instead, what follows here are just a few more anecdotal quotes from teachers of the course about their didactic experiences.

Dora Achourioti (Amsterdam University College). " 'Logic in Action' is not a conventional logic text-book. The conventional introduction to logic would teach the technical material first and then study its applications (if applications are at all meant to be part of the picture). In LiA, logic rarely features in its own, outside of the real life practices where it is most naturally embedded. For the teacher, this presents a challenge. The conventional road is straightforward. But this one is not clear how to follow. How do you make sure to reach the mathematical precision that you do not want to compromise? How do you make sure that by complementing the material on the technical side you do not thereby take away from the richness and breadth of the subject and its various connections with other disciplines as highlighted in the book? It is important to work with a text that allows questions to be asked and teaching to develop, rather than enforcing a rigorous attitude that leaves no room for flexibility and hence no room for improvement. At AUC we have tried to make the most out of this flexibility, using the book more as a rich source of inspiration, rather than a book of instructions on how to do logic."

Wesley Holliday (University of California at Berkeley). "The last time I taught with 'Logic in Action' was in the Spring of 2012 at Stanford University. Nothing from that experience seems in conflict with any of the written impressions in this article. Since then I have used the 'Logic in Action' text as my go-to recommendation for students in my modal logic course and my first-order meta-theory course who want another text for review or to strengthen their understanding. (Incidentally, my text for modal logic is "Modal Logic for Open Minds", and my text for first-order meta-theory last time was Chiswell and Hodges' "Mathematical Logic", plus selections from Enderton, van Dalen, and Hodges' "Elementary Predicate Logic.") Student evaluations have been very positive. My plan is that when I am assigned the introductory logic course at Berkeley (with about 100 students), I will use 'Logic in Action', and then I will have a lot more to say."

Tomohiro Hoshi (Stanford On-Line High School). "The material that 'Logic in Action' provides matches the pedagogical spirit of the Stanford Online High School very well. We believe that active and live engagement of our students is essential for learning processes and have tried to represent this spirit in our online environment. We often feel that some of the most technologically advanced materials with lots of automated support for students do not fit the above goals. By contrast, 'Logic in Action' makes its material accessible to a wide variety of our students, not only by having the text and associated supplementary material free online, but also by grounding technical materials that are often challenging to students of our age groups to fields of study that they can more readily relate to, while still providing great opportunities for our students to experience what we believe is a true learning experience by "getting their hands dirty" with the material they are provided."

Fenrong Liu (Tsinghua University Beijing) "I have used the textbook 'Logic in Action' at Tsinghua, for undergraduate teaching over several years now. It is a one semester course, 48 hours in total. I usually cover propositional logic, first-order logic, epistemic logic and dynamic logic, and sometimes a bit of logic and games. The message of logic as a interdisciplinary subject is well received: this is also confirmed by the structure of the audience, as my students are from mathematics, computer science, philosophy, physics, and engineering. The students found the traditional part of first order logic still rather difficult, but get very excited when we start epistemic logic and dynamic logic. They can easily connect what they learned in class with how they reason with information in real life."

These statements from hands-on teachers largely speak for themselves. Even so, in the following section, I identify some further difficulties as I see them.

8   From plan to reality: difficulties

Self-study is a bridge too far   In the confrontation of this new type of course with reality, several things can be noted. First, there is a big gap between a course taught at institutions by teachers interacting with students and an internet course for pure self-study. Simply put, apart from a few scattered positive reactions from visitors to our website, the second goal does not seem to have been realized at all.

So, let us consider experiences in more standard academic environments. Here, too, much can be noted that leads to serious questions about realizing the intended goals as explained in the above. We list several difficulties, though these do not mean that the 'Logic in Action' spirit is not appreciated!

Emergence of dialects   One thing that is very noticeable is the immediate divergence in the way the course has been used, depending on the teachers' experiences, or their own views on the subject. It seems fair to say that 'Logic in Action' has as many dialects as it has geographical locations. One striking phenomenon is an urge to just add new material to old courses, retaining all the old standard topics such as natural deduction, so that the course loses its more radical character, and rather becomes the addition of, say, some epistemic logic to a relatively standard logic course. Two major emerging dialects that can be discerned are as follows: teach a sequence 'propositional logic, modal logic (in a more formal style), first-order logic', or teach things in the order 'propositional logic, first-order logic, modal logic'.

There may be various reasons for this minimal modus operandi, in whatever order it is done. Perhaps teachers are happy with the standard material in logic introductions, but do not object to adding a few topics to round it out, or make it more up to date. Some teachers have also complained that the course does not provide enough abstract mathematical material and training, which they consider the backbone of logic education: technique first, erudition and broad-mindedness later. Perhaps also, student audiences want more focus, finding the wide spread of topics disorienting. Living in a wide open world is not for everyone.

The target audience   It may also be the case, and this is a perennial issue with introductory logic teaching, that broad audience courses do not work as well as specialized courses catering to the needs, and prejudices, of students from specific disciplines. Moreover, there may still be a perceived bias in our material, despite the intended broad scope. Anecdotal responses to the course have been that it is too much computer science oriented, and too little toward, say, philosophy. Somewhat ironically, a traditional very formal skills-oriented logic course may generate less resistance from specific disciplines. Since these skills do not apply to anything in particular, such a course treats everything equally, having no favorites.

But also, to continue with an earlier observation, university students fall into very different categories. There are 'open-system types' who like flexibility and change, and 'closed-system types' thriving only in well-defined communities with strict norms for what is 'good' and what is not (the 'bad' is often: what is done in other disciplines or paradigms). This division is clearly observable in graduate schools, where one has to cater for the sensitivities of both types of student – and it may well be that a course like this, with its broad range of topics and open windows to the university at large, will rub closed-systems types exactly the wrong way.

Supporting training   But beyond these larger perspectives, there are also simpler down to earth issues with teaching the material in its current form. One difficulty is finding good exercises that test understanding of new topics and new skills. Traditional logic courses have had at least a century of honing test questions in their main fare, while in a new course like this, we need to find a new repertoire to train and test students in understanding the working of questions and other informational acts. This is a serious creative challenge requiring additional investment.

Likewise, our recurrent computational thread raises issues: should not this involve real training in programming or other hands-on computational skills? Some members of our team think so, and therefore, it should be noted that the 'Logic in Action' webpage also has a Part IV, still under development, with concrete programming materials. However, other teachers find this emphasis alienating for students who want a general logic course, not one biased toward computer science.

ICT form or good old content?   As for the intended transformative internet and technology aspects of the course when it started, our main experience has been that this is not a decisive factor in the success of the course as taught. It may be an asset for some students, but given the level of sophistication in the world of education today, a course like this does not offer any creative technology that would give it an advantage over any other. To get ahead in this race, presumably, huge development efforts would be needed. But it is very unclear right now what realistic and desirable goal would justify such an effort (see also Section 9 on this issue).

Practical reasoning and social impact   Finally, returning to the ideal of a self-study logic course with benefits for everyone, it seems clear by now that 'Logic in Action' really functions as an upper-level high school or under-graduate-level university course, and one that is directed mostly toward logic in its academic form. One could also have the goal of improving actual thinking and argumentation practice, but this would require an effort that we have not made. In fact, it is not clear yet if there can be a happy mixture of the abstract intellectual approach in this course and hands-on courses on argumentation theory or critical thinking.

This is not to say that striving for practical impact of logic courses is an endeavor without value. Even more traditional logicians like Evert Willem Beth, a founding father of the Amsterdam *ILLC* environment, explicitly stated his ideal that logic should improve the level of argumentation and reasonable interaction in our society. But realizing such an ambition seems a challenging separate task.

9   Conclusion: where to go from here

We conclude with a few thoughts on where the course 'Logic in Action' might go.

Given the above observations, our current thinking has become more moderate and laissez-faire. The material that we have produced seems a natural and coherent set. Beyond that, it may not be a good idea to impose much ideological uniformity on a course like this, and in any case, enforcing it across the globe is impossible in practice. Moreover, given the fast developments in the academic role of logic, flexibility is needed to accommodate further changes.

The material for a broad logic course in the style described here remains available in the on-line textbook, which will be updated periodically. In addition, we will add teaching tools as they come our way, including course slides and new exercises. We are also thinking of adding a 'best practices forum' where users of the website can meet. Finally, our team is thinking of 'teaching the teachers', offering courses at suitable venues for people considering to use this course.

As for the internet ambitions, it turns out that no harm has been done by procrastinating. The world of education actually seems to have reached a phase where, worldwide, initial expectations about slick on-line courses have been downgraded to much more realistic levels. The current trend toward 'blended learning' (cf. Bersin 2004) emphasizes the indispensable educational role of real teacher-student interactions, over and above what a textbook or website can give. Our course material can help with blended learning, but trying to 'can the course' will not work.

To go further, and find out what really happens with users of our course, we may also consider creating a 'logic garden' – on the analogy of the innovative website 'math garden' created by Han van der Maas at the University of Amsterdam:

http://www.rekentuin.nl/

This would be a site where a wide variety of visitors can experiment with the material presented here, leaving traces that we can use to improve our course, and learn more about what it is to learn logic.

But all these desiderata do not detract from what we see as our main contribution. The course 'Logic in Action' was designed to enrich what students learn in their first encounter with logic. In addition to content of any introduction to a field, there is also a spirit: a modus operandi and even an intellectual value system gets transmitted. Ours has been to make the scope of logic broad, and in line with that, also the students' view of its position in the university arena. While we are not expecting a revolution, given our material and teaching experiences so far, we do believe in the power of small steps in creating large beneficial attitude changes.

References

[1] H. Andréka, J. van Benthem, N. Bezhanishvili & I. Németi, 2014, 'Changing a Semantics: Opportunism or Courage?', in M. Manzano, I. Sain and E. Alonso, eds., *The Life and Work of Leon Henkin*, Birkhaüser Verlag, 307–337.

[2] J. van Benthem, 2011, *Logical Dynamics of Information and Interaction*, Cambridge University Press, Cambridge UK.

[3] J. van Benthem, 2014, *Logic in Games*, The MIT Press, Cambridge MA.

[4] J. Bersin, 2004, *The Blended Learning Book*, Pfeiffer and Company internet publishers.

[5] E. W. Beth, 1963, 'Konstanten van het Wiskundige Denken', *Mededelingen van de Koninklijke Nederlandse Akademie van Wetenschappen*, 26, 231–255.

[6] K. Devlin, 1991, *Logic and Information*, Cambridge University Press, Cambridge UK.

[7] H. van Ditmarsch, W. van der Hoek & B. Kooi, 2007, *Dynamic Epistemic Logic*, Cambridge University Press, Cambridge UK.

[8] A.C. Graham, 2003, *Later Mohist Logic, Ethics and Science*, The Chinese University, Hongkong.

[9] R. Haskell, 2000, *Transfer of Learning: Cognition, Instruction, and Reasoning*, Elsevier Science Publishers, Amsterdam.

    Received 11 October 2016

NaDeA: A Natural Deduction Assistant with a Formalization in Isabelle

Jørgen Villadsen
DTU Compute - Department of Applied Mathematics and Computer Science, Technical University of Denmark, Richard Petersens Plads, Building 324, DK-2800 Kongens Lyngby, Denmark
jovi@dtu.dk

Alexander Birch Jensen
DTU Compute - Department of Applied Mathematics and Computer Science, Technical University of Denmark, Richard Petersens Plads, Building 324, DK-2800 Kongens Lyngby, Denmark

Anders Schlichtkrull
DTU Compute - Department of Applied Mathematics and Computer Science, Technical University of Denmark, Richard Petersens Plads, Building 324, DK-2800 Kongens Lyngby, Denmark

Abstract

We present a new software tool for teaching logic based on natural deduction. Its proof system is formalized in the proof assistant Isabelle such that its definition is very precise. Soundness of the formalization has been proved in Isabelle. The tool is open source software developed in TypeScript / JavaScript and can thus be used directly in a browser without any further installation. Although developed for computer science bachelor students who are used to study and program concrete computer code in a programming language we consider the approach relevant for a broader audience and for other proof systems as well.

Keywords: Natural Deduction, Formalization, Isabelle Proof Assistant, First-Order Logic, Higher-Order Logic.

The present article substantially extends our presentation and tool demonstration at TTL 2015 (Fourth International Conference on Tools for Teaching Logic, Rennes, France, 9–12 June 2015). In particular, a new section with elaboration on the formalization in Isabelle has been added. We would like to thank Stefan Berghofer, Jasmin Christian Blanchette, Mathias Fleury and Dmitriy Traytel for discussions about the formalization of logical inference systems in Isabelle. We would also like to thank Andreas Halkjær From, Andreas Viktor Hess, John Bruntse Larsen, Ashokaditya Mohanty and the referees for comments on the paper.

1   Introduction

In this paper we present the **NaDeA** software tool. First, we provide the motivation and a short description. We then present the natural deduction system as it is done in a popular textbook [15] and as is it done in **NaDeA** by looking at its formalization in Isabelle. This illustrates the differences between the two approaches. We also present the semantics of first-order logic as formalized in Isabelle. Thereafter we explain how **NaDeA** is used to construct a natural deduction proof. After that, we explain the soundness proof of the natural deduction proof system in Isabelle. Lastly, we compare **NaDeA** to other natural deduction assistants and consider how **NaDeA** could be improved.

1.1   Motivation

We have been teaching a bachelor logic course — with logic programming — for a decade using a textbook with emphasis on tableaux and resolution [1]. We have started to use the proof assistant Isabelle [2] and refutation proofs are less preferable here. The proof system of natural deduction [3, 4, 5, 15] with the introduction and elimination rules as well as a discharge mechanism seems more suitable. The natural deduction proof system is widely known, used and studied among logicians throughout the world. However, our experience shows that many of our computer science bachelor students struggle to understand the most difficult aspects.

This also goes for other proof systems. We find that teaching logic to computer science bachelor students can be hard because in our case they do not have a strong theoretical mathematical background. Instead, most students are good at understanding concrete computer code in a programming language. The syntax used in Isabelle is in many ways similar to a programming language. A clear and explicit formalization of first-order logic and a proof system may help the students in understanding important details.

We find it important to teach both the semantics of first-order logic and the soundness proof to bachelor students. In the present course the formal semantics as well as the soundness proof in Isabelle are presented to the students. The formalization is also available online in NaDeA and the entire Isabelle file is available in NaDeA too. However, in the present course the students are not expected to be able to construct such a formalization in Isabelle from scratch.

The proof assistant Isabelle is different from a programming language because the expressions are not necessarily computable. For instance, quantifications over infinite domains are possible.

1.2 The Tool

We present the natural deduction assistant **NaDeA** with a formalization of its proof system in the proof assistant Isabelle. It can be used directly in a browser without any further installation and is available here:

http://nadea.compute.dtu.dk/

NaDeA is open source software developed in TypeScript / JavaScript and stored on GitHub. The formalization of its proof system in Isabelle is available here:

http://logic-tools.github.io/

Once **NaDeA** is loaded in the browser — about 250 KB with the jQuery Core library — no internet connection is required. Therefore **NaDeA** can also be stored locally.

We present the proof in an explicit code format that is equivalent to the Isabelle syntax, but with a few syntactic differences to make it easier to understand for someone trying to learn Isabelle. In this format, we present the proof in a style similar to that of Fitch's diagram proofs. We avoid the seemingly popular Gentzen's tree style to focus less on a visually pleasing graphical representation that is presumably much more challenging to implement.

We find that the following requirements constitute the key ideals for any natural deduction assistant. It should be:

– Easy to use.

– Clear and explicit in every detail of the proof.

– Based on a formalization that can be proved at least sound, but preferably also complete.

Based on this, we saw an opportunity to develop **NaDeA** which offers help for new users, but also serves to present an approach that is relevant to the advanced users.

In a paper considering the tools developed for teaching logic over the last decade [14, p. 137], the following is said about assistants (not proof assistants like Isabelle but tools for learning/teaching logic):

> Assistants are characterized by a higher degree of interactivity with the user. They provide menus and dialogues to the user for interaction purposes. This kind of tool gives the students the feeling that they are being helped in building the solution. They provide error messages and hints

in the guidance to the construction of the answer. Many of them usually offer construction of solution in natural deduction proofs. [...] They are usually free licensed and of open access.

We think that this characterization in many ways fits NaDeA. While NaDeA might not bring something new to the table in the form of delicate graphical features, we emphasize the fact that it has some rather unique features such as a formalization of its proof system in Isabelle.

2 Natural Deduction in a Textbook

We consider natural deduction as presented in a popular textbook on logic in computer science [15]. First, we take a look at substitution, which is central to the treatment of quantifiers in natural deduction.

2.1 On Substitution

The following definition for substitution is used in [15, p. 105 top]:

> Given a variable x, a term t and a formula ϕ we define $\phi[t/x]$ to be the formula obtained by replacing each free occurrence of variable x in ϕ with t.

The usual side conditions that come with rules using this substitution seem to be omitted, but we are shortly after [15, p. 106 top] given the following definition of what it means that 't must be free for x in ϕ':

> Given a term t, a variable x and a formula ϕ, we say that t is free for x in ϕ if no free x leaf in ϕ occurs in the scope of $\forall y$ or $\exists y$ for any variable y occurring in t.

The following quote [15, p. 106 bottom] emphasizes the side conditions:

> It might be helpful to compare 't is free for x in ϕ' with a precondition of calling a procedure for substitution. If you are asked to compute $\phi[t/x]$ in your exercises or exams, then that is what you should do; but any reasonable implementation of substitution used in a theorem prover would have to check whether t is free for x in ϕ and, if not, rename some variables with fresh ones to avoid the undesirable capture of variables.

In our formalization such notions and their complications become easier to explain because all side conditions of the rules are very explicitly stated. We see it as one of the major advantages of presenting this formalization to students.

2.2 Natural Deduction Rules

We now present the natural deduction rules as described in the literature, again using [15]. The first 9 are rules for classical propositional logic and the last 4 are for first-order logic. Intuitionistic logic can be obtained by omitting the rule *PBC* (proof by contradiction, called "Boole" later) and adding the $\bot$-elimination rule (also known as the rule of explosion) [16]. The rules are as follows:

$$\frac{\begin{array}{c}\boxed{\begin{array}{c}\neg\phi\\\vdots\\\bot\end{array}}\end{array}}{\phi}\,PBC \qquad \frac{\phi \quad \phi\rightarrow\psi}{\psi}\,\rightarrow E \qquad \frac{\boxed{\begin{array}{c}\phi\\\vdots\\\psi\end{array}}}{\phi\rightarrow\psi}\,\rightarrow I$$

$$\frac{\phi\vee\psi \quad \boxed{\begin{array}{c}\phi\\\vdots\\\chi\end{array}} \quad \boxed{\begin{array}{c}\psi\\\vdots\\\chi\end{array}}}{\chi}\,\vee E \qquad \frac{\phi}{\phi\vee\psi}\,\vee I_1 \qquad \frac{\psi}{\phi\vee\psi}\,\vee I_2$$

$$\frac{\phi\wedge\psi}{\phi}\,\wedge E_1 \qquad \frac{\phi\wedge\psi}{\psi}\,\wedge E_2 \qquad \frac{\phi \quad \psi}{\phi\wedge\psi}\,\wedge I$$

$$\frac{\exists x\,\phi \quad \boxed{\begin{array}{cc}x_0 & \phi\,[x_0/x]\\ & \vdots\\ & \chi\end{array}}}{\chi}\,\exists E \qquad \frac{\phi\,[t/x]}{\exists x\,\phi}\,\exists I$$

$$\frac{\forall x\,\phi}{\phi\,[t/x]}\,\forall E \qquad \frac{\boxed{\begin{array}{c}x_0\\\vdots\\\phi\,[x_0/x]\end{array}}}{\forall x\,\phi}\,\forall I$$

Side conditions to rules for quantifiers:

$\exists E$: x_0 does not occur outside its box (and therefore not in χ).

$\exists I$: t must be free for x in ϕ.

$\forall E$: t must be free for x in ϕ.

$\forall I$: x_0 is a new variable which does not occur outside its box.

In addition there is a special copy rule [15, p. 20]:

> A final rule is required in order to allow us to conclude a box with a formula which has already appeared earlier in the proof. [...] The copy rule entitles us to copy formulas that appeared before, unless they depend on temporary assumptions whose box has already been closed.

The copy rule is not needed in our formalization due to the way it manages a list of assumptions.

As it can be seen, there are no rules for truth, negation or biimplication, but the following equivalences can be used:

$$\begin{aligned} \top &\equiv \bot \to \bot \\ \neg A &\equiv A \to \bot \\ A \leftrightarrow B &\equiv (A \to B) \wedge (B \to A) \end{aligned}$$

The symbols A and B are arbitrary formulas.

3   Natural Deduction in NaDeA

One of the unique features of NaDeA is that it comes with a formalization in Isabelle of the natural deduction proof system, including a proof in Isabelle of the soundness theorem for the proof system. In this section we present the definitions necessary for expressing the soundness theorem and the proof in Isabelle is presented in section 5.

3.1   Syntax for Terms and Formulas

The terms and formulas of the first-order logic language are defined as the datatypes term and formula (later abbreviated tm and fm, respectively). The type identifier represents predicate and function symbols (later abbreviated id).

identifier := string

term := Var nat | Fun identifier [term, ..., term]

formula := Falsity | Pre identifier [term, ..., term] | Imp formula formula |
           Dis formula formula | Con formula formula |
           Exi formula | Uni formula

Truth, negation and biimplication are abbreviations. In the syntax of our formalization, we refer to variables by use of the de Bruijn indices. That is, instead of identifying a variable by use of a name, usually x, y, z etc., each variable has an index that determines its scope. The use of de Bruijn indices instead of named variables allows for a simple definition of substitution. Furthermore, it also serves the purpose of teaching the students about de Bruijn indices. Note that we are not advocating that de Bruijn indices replace the standard treatment of variables in general. It arguably makes complex formulas harder to read, but the pedagogical advantage is that the notion of scope is practiced.

3.2 Natural Deduction Rules

Provability in NaDeA is defined inductively as follows (OK p z means that the formula p follows from the list of assumptions z and member p z means that p is a member of the list z):

$$\frac{\text{member p z}}{\text{OK p z}} \text{ Assume} \qquad \frac{\text{OK Falsity ((Imp p Falsity) \# z)}}{\text{OK p z}} \text{ Boole}$$

$$\frac{\text{OK (Imp p q) z} \qquad \text{OK p z}}{\text{OK q z}} \text{ Imp\_E} \qquad \frac{\text{OK q (p \# z)}}{\text{OK (Imp p q) z}} \text{ Imp\_I}$$

$$\frac{\text{OK (Dis p q) z} \qquad \text{OK r (p \# z)} \qquad \text{OK r (q \# z)}}{\text{OK r z}} \text{ Dis\_E}$$

$$\frac{\text{OK p z}}{\text{OK (Dis p q) z}} \text{ Dis\_I1} \qquad \frac{\text{OK q z}}{\text{OK (Dis p q) z}} \text{ Dis\_I2}$$

$$\frac{\text{OK (Con p q) z}}{\text{OK p z}} \text{ Con\_E1} \qquad \frac{\text{OK (Con p q) z}}{\text{OK q z}} \text{ Con\_E2} \qquad \frac{\text{OK p z} \qquad \text{OK q z}}{\text{OK (Con p q) z}} \text{ Con\_I}$$

$$\frac{\text{OK (Exi p) z} \qquad \text{OK q ((sub 0 (Fun c []) p) \# z)} \qquad \text{news c (p\#q\#z)}}{\text{OK q z}} \text{ Exi\_E}$$

$$\frac{\text{OK (sub 0 t p) z}}{\text{OK (Exi p) z}} \text{ Exi\_I}$$

$$\frac{\text{OK (Uni p) z}}{\text{OK (sub 0 t p) z}} \text{ Uni\_E} \qquad \frac{\text{OK (sub 0 (Fun c []) p) z} \qquad \text{news c (p \# z))}}{\text{OK (Uni p) z}} \text{ Uni\_I}$$

Instead of writing OK p z we could also use the syntax z ⊢ p, even in Isabelle, but we prefer a more programming-like approach.

The operator # is between the head and the tail of a list. news c l checks if the identifier c does not occur in any of the formulas in the list l and sub n t p returns the formula p where the term t has been substituted for the variable with the de Bruijn index n.

Note that new constants instead of variables not occuring in the assumptions are used in the existential elimination rule and in the universal introduction rule.

In the types we use ⇒ for function spaces. We include the definitions of member, news and sub because they are necessary for the soundness theorem and also for the formalization in section 5:

```
member :: fm ⇒ fm list ⇒ bool

member p [] = False
member p (q # z) = (if p = q then True else member p z)


new_term :: id ⇒ tm ⇒ bool

new_term c (Var n) = True
new_term c (Fun i l) = (if i = c then False else new_list c l)

new_list :: id ⇒ tm list ⇒ bool

new_list c [] = True
new_list c (t # l) = (if new_term c t then new_list c l else False)


new :: id ⇒ fm ⇒ bool

new c Falsity = True
new c (Pre i l) = new_list c l
new c (Imp p q) = (if new c p then new c q else False)
new c (Dis  p q) = (if new c p then new c q else False)
new c (Con p q) = (if new c p then new c q else False)
new c (Exi  p) = new c p
new c (Uni p) = new c p

news :: id ⇒ fm list ⇒ bool

news c [] = True
news c (p # z) = (if new c p then news c z else False)
```

```
inc_term :: tm ⇒ tm

inc_term (Var n) = Var (n + 1)
inc_term (Fun i l) = Fun i (inc_list l)

inc_list :: tm list ⇒ tm list

inc_list [] = []
inc_list (t # l) = inc_term t # inc_list l


sub_term :: nat ⇒ tm ⇒ tm ⇒ tm

sub_term v s (Var n) = (if n < v then Var n else if n = v then s else Var (n − 1))
sub_term v s (Fun i l) = Fun i (sub_list v s l)

sub_list :: nat ⇒ tm ⇒ tm list ⇒ tm list

sub_list v s [] = []
sub_list v s (t # l) = sub_term v s t # sub_list v s l


sub :: nat ⇒ tm ⇒ fm ⇒ fm

sub v s Falsity = Falsity
sub v s (Pre i l) = Pre i (sub_list v s l)
sub v s (Imp p q) = Imp (sub v s p) (sub v s q)
sub v s (Dis  p q) = Dis  (sub v s p) (sub v s q)
sub v s (Con p q) = Con (sub v s p) (sub v s q)
sub v s (Exi p) = Exi (sub (v + 1) (inc_term s) p)
sub v s (Uni p) = Uni (sub (v + 1) (inc_term s) p)
```

3.3   Semantics for Terms and Formulas

To give meaning to formulas and to prove NaDeA sound we need a semantics of the first-order logic language. We present the semantics below. e is the environment, i.e. a mapping of variables to elements. f maps function symbols to the maps they represent. These maps are from lists of elements of the universe to elements of the universe. Likewise, g maps predicate symbols to the maps they represent. 'a is a type variable that represents the universe. It can be instantiated with any type. For instance, it can be instantiated with the natural numbers, the real numbers or strings.

semantics_term :: (nat $\Rightarrow$ 'a) $\Rightarrow$ (id $\Rightarrow$ 'a list $\Rightarrow$ 'a) $\Rightarrow$ tm $\Rightarrow$ 'a

semantics_term e f (Var n) = e n
semantics_term e f (Fun i l) = f i (semantics_list e f l)

semantics_list :: (nat $\Rightarrow$ 'a) $\Rightarrow$ (id $\Rightarrow$ 'a list $\Rightarrow$ 'a) $\Rightarrow$ tm list $\Rightarrow$ 'a list

semantics_list e f [] = []
semantics_list e f (t # l) = semantics_term e f t # semantics_list e f l

semantics :: (nat $\Rightarrow$ 'a) $\Rightarrow$ (id $\Rightarrow$ 'a list $\Rightarrow$ 'a) $\Rightarrow$ (id $\Rightarrow$ 'a list $\Rightarrow$ bool) $\Rightarrow$
$\qquad\qquad\qquad\qquad\qquad\qquad\qquad\qquad\qquad\qquad\qquad$ fm $\Rightarrow$ bool

semantics e f g Falsity = False
semantics e f g (Pre i l) = g i (semantics_list e f l)
semantics e f g (Imp p q) = (if semantics e f g p then semantics e f g q else True)
semantics e f g (Dis p q) = (if semantics e f g p then True else semantics e f g q)
semantics e f g (Con p q) = (if semantics e f g p then semantics e f g q else False)
semantics e f g (Exi p) =
$\qquad\qquad\qquad$ (? x. semantics (% n. if n = 0 then x else e (n − 1)) f g p)
semantics e f g (Uni p) =
$\qquad\qquad\qquad$ (! x. semantics (% n. if n = 0 then x else e (n − 1)) f g p)

Most of the cases of **semantics** should be self-explanatory, but the Uni case is complicated. The details are not important here, but in the case for Uni it uses the universal quantifier (!) of Isabelle's higher-order logic to consider all values of the universe. It also uses the lambda abstraction operator (%) to keep track of the indices of the variables. Likewise, the case for Exi uses the existential quantifier (?) of Isabelle's higher-order logic.

We have proved soundness of the formalization in Isabelle (shown here as a derived rule):

$$\frac{\text{OK p []}}{\text{semantics e f g p}} \text{ Soundness}$$

This result makes **NaDeA** interesting to a broader audience since it gives confidence in the formulas proved using the tool. The proof in Isabelle of the soundness theorem is presented in section 5.

4   Construction of a Proof

We show here how to build and edit proofs in **NaDeA**. Furthermore, we describe the presentation of proofs in **NaDeA**.

In order to start a proof, you have to start by specifying the goal formula, that is, the formula you wish to prove. To do so, you must enable editing mode by clicking the Edit button in the top menu bar. This will show the underlying proof code and you can build formulas by clicking the red ¤ symbol. Alternatively, you can load a number of tests by clicking the Load button.

At all times, once you have fully specified the conclusion of any given rule, you can continue the proof by selecting the next rule to apply. Again you can do this by clicking the red ¤ symbol. Furthermore, NaDeA allows for undoing and redoing editing steps with no limits.

All proofs are conducted in backward-chaining mode. That is, you must start by specifying the formula that you wish to prove. You then apply the rules inductively until you reach a proof — if you can find one. The proof is finished by automatic application of the Assume rule once the conclusion of a rule is found in the list of assumptions.

To start over on a new proof, you can load the blank proof by using the Load button, or you can refresh the page.

In NaDeA we present any given natural deduction proof (or an attempt at one) in two different types of syntax. One syntax follows the rules as defined in section 3 and is closely related to the formalization in Isabelle, but with a simplified syntax that makes it suitable for teaching purposes. The proof is not built as most often seen in the literature about natural deduction. Usually, for each rule the premises are placed above its conclusion separated by a line. We instead follow the procedure of placing each premise of the rule on separate lines below its conclusion with an additional level of indentation. Here is a screenshot followed by the proof tree:

Natural Deduction Assistant

1	Imp_I	[] $P \wedge (P \rightarrow Q) \rightarrow Q$
2	Imp_E	$[P \wedge (P \rightarrow Q)]$ Q
3	Con_E2	$[P \wedge (P \rightarrow Q)]$ $P \rightarrow Q$
4	Assume	$[P \wedge (P \rightarrow Q)]$ $P \wedge (P \rightarrow Q)$
5	Con_E1	$[P \wedge (P \rightarrow Q)]$ P
6	Assume	$[P \wedge (P \rightarrow Q)]$ $P \wedge (P \rightarrow Q)$

$$\cfrac{\cfrac{\overline{\rule{2cm}{0.4pt}}}{\cfrac{p \wedge (p \to q)}{p \to q}}{}^{(1)} \quad \cfrac{\overline{\rule{2cm}{0.4pt}}}{\cfrac{p \wedge (p \to q)}{p}}{}^{(1)}}{\cfrac{q}{p \wedge (p \to q) \to q}}{}^{(1)}$$

The above proof can also be written in terms of the **OK** syntax as follows:

1 OK (Imp (Con (Pre "P" []) (Imp (Pre "P" []) (Pre "Q" []))) (Pre "Q" [])) [] Imp_I

2 OK (Pre "Q" []) [(Con (Pre "P" []) (Imp (Pre "P" []) (Pre "Q" [])))] Imp_E

3 OK (Imp (Pre "P" []) (Pre "Q" []))
   [(Con (Pre "P" []) (Imp (Pre "P" []) (Pre "Q" [])))] Con_E2

4 OK (Con (Pre "P" []) (Imp (Pre "P" []) (Pre "Q" [])))
   [(Con (Pre "P" []) (Imp (Pre "P" []) (Pre "Q" [])))] Assume

5 OK (Pre "P" []) [Con (Pre "P" []) (Imp (Pre "P" []) (Pre "Q" []))] Con_E1

6 OK (Con (Pre "P" []) (Imp (Pre "P" []) (Pre "Q" [])))
   [(Con (Pre "P" []) (Imp (Pre "P" []) (Pre "Q" [])))] Assume

So in a way we have the two presentation styles. However, the standard form displayed in the screenshot is always presented and the programming style with the **OK** syntax is switched on and off with a single click in the browser. The programming style is mandatory when a formula must be entered. We find that the students in general prefer the standard form but also that the switch to the programming style when necessary is rather unproblematic.

For a small but quite interesting example of a proof of a first-order formula consider the following screenshot:

1	Imp_I	[ ] $(\forall x.P(x)) \vee (\forall x.Q(x)) \to (\forall x.P(x) \vee Q(x))$
2	Uni_I	$[(\forall x.P(x)) \vee (\forall x.Q(x))]$ $\forall x.P(x) \vee Q(x)$
3	Dis_E	$[(\forall x.P(x)) \vee (\forall x.Q(x))]$ $P(c') \vee Q(c')$
4	Assume	$[(\forall x.P(x)) \vee (\forall x.Q(x))]$ $(\forall x.P(x)) \vee (\forall x.Q(x))$
5	Dis_I1	$[\forall x.P(x), (\forall x.P(x)) \vee (\forall x.Q(x))]$ $P(c') \vee Q(c')$
6	Uni_E	$[\forall x.P(x), (\forall x.P(x)) \vee (\forall x.Q(x))]$ $P(c')$
7	Assume	$[\forall x.P(x), (\forall x.P(x)) \vee (\forall x.Q(x))]$ $\forall x.P(x)$
8	Dis_I2	$[\forall x.Q(x), (\forall x.P(x)) \vee (\forall x.Q(x))]$ $P(c') \vee Q(c')$
9	Uni_E	$[\forall x.Q(x), (\forall x.P(x)) \vee (\forall x.Q(x))]$ $Q(c')$
10	Assume	$[\forall x.Q(x), (\forall x.P(x)) \vee (\forall x.Q(x))]$ $\forall x.Q(x)$
11	*	

66

The line with the * in the proof is for the side condition that requires that the constant c′ is new. By clicking on the proof the check is displayed in the OK syntax as follows:

news (Fun "c*" []) [(Uni (Dis (Pre "P" [Var 0]) (Pre "Q" [Var 0]))),
          (Dis (Uni (Pre "P" [Var 0])) (Uni (Pre "Q" [Var 0]))))]

The constant c′ is written as "c*" here.

5 Formalization in Isabelle

Formalizations in Isabelle are written in a language that combines functional programming and logic. Our computer science bachelor students know programming from an introductory programming course and are introduced to logic in our course. This makes Isabelle a well suited way to present a sound proof system compared to a more abstract and mathematical approach. Furthermore, the language used in Isabelle is somewhat close to English, which also aids the intuitions of the students. Isabelle also allows the students to interactively inspect the different states of the proof and get an overview of the lemmas and theorems that are used in the steps – all in one screen. In this section we present the soundness proof using our formalization and show the concepts known from programming and logic.

5.1 An Overview

We first give an overview of the formalization in Isabelle. In the overview we see a number of datatypes *tm* and *fm*, that represent the objects that we want to reason about. We also see a primitive recursive function *member* which is used in the inductive definition *OK* of the proof system. Lastly, we see the *soundness* theorem of the proof system. We will explain these concepts as well as show and elaborate on the parts of the formalization that we did not put in the overview.

theory *NaDeA* ***imports*** *Main* ***begin***

type_synonym *id = "char list"*

datatype *tm = Var nat | Fun id "tm list"*

datatype *fm = Falsity | Pre id "tm list" | Imp fm fm | Dis fm fm | Con fm fm |*
          Exi fm | Uni fm

primrec
  member :: "fm ⇒ fm list ⇒ bool"
where
  "member p [] = False" |
  "member p (q # z) = (**if** p = q **then** True **else** member p z)"

(* More primitive recursive functions as included in the previous sections *)

inductive
  OK :: "fm ⇒ fm list ⇒ bool"
where
Assume:
      "member p z ⟹ OK p z" |
Boole:
      "OK Falsity ((Imp p Falsity) # z) ⟹ OK p z" |
Imp_E:
      "OK (Imp p q) z ⟹ OK p z ⟹ OK q z" |
Imp_I:
      "OK q (p # z) ⟹ OK (Imp p q) z" |

(* More rules as included in the previous sections *)

(* A proof of soundness' is included in the following sections *)

theorem soundness: "OK p [] ⟹ semantics e f g p"
  proof (simp add: soundness') **qed**

end

5.2   Terms and Formulas

Terms are defined by a datatype *tm*. Datatypes are a well-known concept from functional programming. A term is either a variable or a function application. Therefore, we have a constructor *Var* which constructs a variable from a *nat* representing its de Bruijn index. Likewise, we have a constructor *Fun* which constructs a function application from an *id* which is its function identifier and a *"tm list"* which represents its subterms.

When we introduce a datatype in Isabelle, we implicitly state that all terms can be constructed from its constructors. We also implicitly state that if two terms are equal then they must have been constructed from the same constructor and arguments. [18]

Formulas are also formalized as a datatype *fm*. It has a constructor for each operator and quantifier of our first-order logic.

5.3   Membership and Other Primitive Recursive Functions

List membership is defined as a primitive recursive function *member* over lists. The constructor for lists is $\#$ which separates the head of the list from the tail. The *member* function is primitive recursive because it removes a constructor from one of its arguments in every recursive call [2]. In Isabelle, primitive recursive functions are defined in much the same way as in functional programming, namely by stating cases for the different constructors.

The intuition of the function is that *member p z* returns true if the formula *p* is found in the list of formulas *z* and false otherwise. The function considers two cases: either the list is empty or it has a head and a tail. In the first case it is clear that the formula is not a member of the list. In the second case, we use the pattern *(q # z)* where *q* is the head of the list and *z* is the tail. If the head is equal to *p* it is true that *p* is a member of the list. Otherwise, we continue by looking in the tail of the list.

Other primitive recursive functions used in the theory are *semantics_term*, *semantics_list*, *semantics*, *new*, *news*, *inc_term*, *inc_list*, *sub_term*, *sub_list* and *sub*. These functions define the semantics, increasing the de Bruijn indices of a term, a constant being new to an expression, and substitution.

5.4   Proof System

Our proof system is defined by an inductive predicate. Each of the rules of the system is a case in the inductive predicate. For instance, consider the following rule:

Assume: "member p z $\implies$ OK p z"

The rule means that *OK p z* follows from *member p z*. Another case is the more complex rule:

Imp_E: "OK (Imp p q) z $\implies$ OK p z $\implies$ OK q z"

It states that *OK q z* follows from *OK (Imp p q) z* and *OK p z*. This corresponds to the usual notation for inference rules:

$$\frac{\text{OK (Imp p q) z} \qquad \text{OK p z}}{\text{OK q z}} \; \text{Imp\_E}$$

That a predicate is inductive means that it holds exactly when it can be derived using the given cases.

5.5   Proof of Soundness

We are now ready for the proof of soundness.

fun
   $put :: "(nat \Rightarrow 'a) \Rightarrow nat \Rightarrow 'a \Rightarrow nat \Rightarrow 'a"$
where
   $"put \ e \ v \ x = (\lambda n. \ \textbf{if} \ n < v \ \textbf{then} \ e \ n \ \textbf{else} \ \textbf{if} \ n = v \ \textbf{then} \ x \ \textbf{else} \ e \ (n - 1))"$

The function *put* updates an environment by mapping variable v to value x. This is used in the definition of the quantifiers, but always for the outermost bound variable. Existing variables greater than v are pushed one position up, i.e. variable i now points to the value of variable $i - 1$ in the old environment.

We use **fun** to declare many different functions without being restricted to the primitive recursive form. The operator λ is for lambda abstraction applied to occurrences of the parameter value and is known from functional programming. More informally, if E is some expression in Isabelle then $\lambda x. \ E \ x$ is the function that takes an input, for instance y, and returns $E \ y$.

lemma $"put \ e \ 0 \ x = (\lambda n. \ \textbf{if} \ n = 0 \ \textbf{then} \ x \ \textbf{else} \ e \ (n - 1))" \ \textbf{proof} \ simp \ \textbf{qed}$

This lemma shows that *put* is a generalization of the expression

$$\lambda n. \ \textbf{if} \ n = 0 \ \textbf{then} \ x \ \textbf{else} \ e \ (n - 1)$$

which appears in the semantics. We use this generalization to prove properties of putting that we use in our soundness proof. The lemma is followed by a proof. In this case, the proof is performed automatically by the simplifier *simp*. The beginning of the proof is marked by **proof** and the end is marked by **qed**. The proof method *simp* works by applying simplification rules [18]. It contains rules that are generated from definitions of functions, datatypes, etc., in addition to simplification rules from the Isabelle library.

lemma *increment:*
   $"semantics\_term \ (put \ e \ 0 \ x) \ f \ (inc\_term \ t) = semantics\_term \ e \ f \ t"$
   $"semantics\_list \ (put \ e \ 0 \ x) \ f \ (inc\_list \ l) = semantics\_list \ e \ f \ l"$
proof *(induct t **and** l rule: semantics_term.induct semantics_list.induct)*
qed *simp_all*

The lemma *increment* shows that we preserve the semantics of a term when we increment its de Bruijn indices while putting a value x at index 0. The reason is that putting pushes the values one index up in the environment. The proof is by induction on t and l, which is stated as

*induct t **and** l rule: semantics_term.induct semantics_list.induct* and it generates four proof goals; one for each of the cases in *semantics_term* and *semantics_list*. These goals can be inspected in the Isabelle editor by placing the cursor right after

> *(induct t **and** l rule: semantics_term.induct semantics_list.induct)*

and looking in the so-called state panel. The proof method *simp_all* applies the simplifier to all available proof goals [18]. We place *simp_all* after **qed** in order to finish the proof and to allow inspection of the proof state interactively in Isabelle.

lemma *commute:* *"put (put e v x) 0 y = put (put e 0 y) (v + 1) x"*
proof *force* **qed**

The lemma *commute* shows that the function *put* commutes. More precisely, we want to put a value at position $v + 1$ in the environment and one at position 0, and the theorem shows that the order in which we do this does not matter, as long as we are careful with the indices.

The proof is automatic and uses the proof method *force*, which works by simplification and classical reasoning [2].

fun
   all :: *"(fm ⇒ bool) ⇒ fm list ⇒ bool"*
where
   *"all b z = (∀p. **if** member p z **then** b p **else** True)"*

The function *all* checks if the predicate *b* is true for all formulas in a list. The ∀ operator is for universal quantification.

lemma *allhead:* *"all b (p # z) ⟹ b p"* **proof** *simp* **qed**

lemma *alltail:* *"all b (p # z) ⟹ all b z"* **proof** *simp* **qed**

lemma *allnew:* *"all (new c) z = news c z"*
proof *(induct z)* **qed** *(simp, simp, metis)*

The lemma *allhead* states that if *b* holds for the entire list, then it holds for the head of the list in particular. The lemma *alltail* is similar, but for the tail of the list. Finally, the lemma *allnew* shows the equivalence between *news* and *all* combined with *new*. The proof uses the proof methods *simp* and *metis* in the order they are written, i.e. *simp* the first proof goal generated by the structural induction on *z*. Then *simp* simplifies the second proof goal which is afterwards proved by *metis*. The *metis* proof method is a resolution theorem prover [17].

71

lemma *map':*
  "new_term c t $\Longrightarrow$ semantics_term e (f(c := m)) t = semantics_term e f t"
  "new_list c l $\Longrightarrow$ semantics_list e (f(c := m)) l = semantics_list e f l"
proof *(induct t **and** l rule: semantics_term.induct semantics_list.induct)*
qed *(simp, simp, metis, simp, simp, metis)*

lemma *map:*
  "new c p $\Longrightarrow$ semantics e (f(c := m)) g p = semantics e f g p"
proof *(induct p arbitrary: e)*
qed *(simp, simp, metis map'(2), simp, metis, simp, metis, simp, metis, simp_all)*

lemma *allmap:*
  "news c z $\Longrightarrow$ all (semantics e (f(c := m)) g) z = all (semantics e f g) z"
proof *(induct z)* **qed** *(simp, simp, metis map)*

The lemma *map'* shows that we preserve the semantics of a term if we map a constant that is new to the term to another value. Here, *f(c := m)* maps function identifier *c* to *m* in the function map *f*. Because the lemma is quite obvious it can be proved automatically. The first and third goals are proved by *simp*, and the second and fourth are simplified by *simp* and then proved by *metis*. The lemma *map* shows that the property of *map'* can be extended to also hold for formulas. This can also be proved automatically. There are seven proof goals of the induction corresponding to each of the formula constructors. We use *simp* to discharge of the first proof goal, then *simp* followed by *metis* for the next four. This time we use *metis map'(2)* to prove the case for predicates. This works by applying *metis* with the addition of the second part of *map'* as a fact with which it can reason. The last two proof goals are proved with the simplifier using *simp_all*. The lemma *allmap* further extends the property of the lemma *map'* to also hold for lists of formulas. We prove it using *simp* and *metis map*.

lemma *substitute':*
  "semantics_term e f (sub_term v s t) =
    semantics_term (put e v (semantics_term e f s)) f t"
  "semantics_list e f (sub_list v s l) =
    semantics_list (put e v (semantics_term e f s)) f l"
proof *(induct t **and** l rule: semantics_term.induct semantics_list.induct)*
qed *simp_all*

The lemma *substitute'* is the famous substitution lemma for terms. This lemma shows a relation between the world of syntax and the world of semantics. More specifically, the relation is between the syntactical operation of substitution and the semantic notion of variable environments. The two are related because a substitution instantiates a variable with a term, and this term represents a value. Thus we get the same semantics of the term if we instead of substitution put the value directly at the index of the variable in the environment. The proof is by induction and *simp_all*.

lemma *substitute:*
  "semantics e f g (sub v t p) = semantics (put e v (semantics_term e f t)) f g p"
proof *(induct p arbitrary: e v t)*
  fix *i l e v t*
  show *"semantics e f g (sub v t (Pre i l)) =*
      semantics (put e v (semantics_term e f t)) f g (Pre i l)"
  proof *(simp add: substitute'(2))* **qed**
next
  fix *p e v t* **assume** *∗: "semantics e' f g (sub v' t' p) =*
      semantics (put e' v' (semantics_term e' f t')) f g p" **for** *e' v' t'*
  have *"semantics e f g (sub v t (Exi p)) =*
      (∃x. semantics (put (put e 0 x) (v + 1)
        (semantics_term (put e 0 x) f (inc_term t))) f g p)"
    using *∗* **proof** *simp* **qed**
  also have *"... =*
      (∃x. semantics (put (put e v (semantics_term e f t)) 0 x) f g p)"
    using *commute increment(1)* **proof** *metis* **qed**
  finally show *"semantics e f g (sub v t (Exi p)) =*
      semantics (put e v (semantics_term e f t)) f g (Exi p)" **proof** *simp* **qed**
  have *"semantics e f g (sub v t (Uni p)) =*
      (∀x. semantics (put (put e 0 x) (v + 1)
        (semantics_term (put e 0 x) f (inc_term t))) f g p)"
    using *∗* **proof** *simp* **qed**
  also have *"... =*
      (∀x. semantics (put (put e v (semantics_term e f t)) 0 x) f g p)"
    using *commute increment(1)* **proof** *metis* **qed**
  finally show *"semantics e f g (sub v t (Uni p)) =*
      semantics (put e v (semantics_term e f t)) f g (Uni p)" **proof** *simp* **qed**
qed *simp_all*

The lemma *substitute* extends the substitution lemma to hold also for formulas. The proof is by induction on a formula p. In the proof we write *arbitrary: e v t* because then e, v and t are also arbitrary in the induction hypothesis. This more general induction hypothesis is necessary for the proof. Most cases can be proven by the simplifier without any instructions, but we prove the cases for predicates *Pre*,

73

existential quantification *Exi* and universal quantification *Uni* more explicitly. For the predicates, we only need instruct the simplifier to use *substitute'(2)* as a simplification rule by writing *(simp add: substitute'(2))*. For the existential quantification we make an explicit proof. We fix the subformula *p* of an existential quantification for which we want to prove the property. As said, we want to prove it with an arbitrary variable environment *e*, an arbitrary variable *v*, and an arbitrary term *t* so we fix those as well. We then state the induction hypothesis ∗ which says that for the subformula *p* of our existential quantification we can put the value of the term *t* in the environment instead of doing substitution with *t*:

> **assume** ∗: "*semantics e' f g (sub v' t' p)* =
> *semantics (put e' v' (semantics_term e' f t')) f g p*" **for** *e' v' t'*

The **for** keyword ensures that *e'*, *v'*, and *t'* are arbitrary as we wished. We wish to prove the substitution lemma for the existential quantification *Exi p*, i.e. that

> *semantics e f g (sub v t (Exi p))* =
> *semantics (put e v (semantics_term e f t)) f g (Exi p)*

The keyword **also** together with **finally** is used to make a proof from left to right of the equality of two expressions. This is what we want to do, and thus we start from the left-hand side:

> *semantics e f g (sub v t (Exi p))*

and realize that by the definition of substitution and the semantics of *Exi* we just need a single value *x* for which the semantics of *sub (v + 1) (inc_term t) p* is true under the environment *put e 0 x*. At the same time, we realize that we can now use the induction hypothesis. Therefore, instead of considering the semantics of *sub (v + 1) (inc_term t) p* under *put e 0 x*, we equivalently consider the semantics of *p* under the variable environment which is *put e 0 x* with the value of *t* put on index *v* + 1. We must thus continue our proof from

> *(∃x. semantics (put (put e 0 x) (v + 1)*
> *(semantics_term (put e 0 x) f (inc_term t))) f g p)*

We can make this expression much simpler by using *commute* and *increment(1)*.

> *(∃x. semantics (put (put e v (semantics_term e f t)) 0 x) f g p)*

We finish our proof using the semantics of *Exi*, as well as the fact that *put* generalizes putting at index 0, and we get the right-hand side we were looking for:

> *semantics (put e v (semantics_term e f t)) f g (Exi p)*

Then follows a proof of substitution for the universal quantification *Uni* since it has the same induction hypothesis. The proof is very similar. Finally we write **qed** *simp_all* to prove the remaining cases by simplification.

lemma *soundness': "OK p z ⟹ all (semantics e f g) z ⟹ semantics e f g p"*
proof *(induct arbitrary: f rule: OK.induct)*
  fix *f p z* **assume** *"all (semantics e f g) z"*
    "all (semantics e f' g) (Imp p Falsity # z) ⟹
       semantics e f' g Falsity" **for** *f'*
  then show *"semantics e f g p"* **proof** *force* **qed**
next
  fix *f p q z r* **assume** *"all (semantics e f g) z"*
    "all (semantics e f' g) z ⟹ semantics e f' g (Dis p q)"
    "all (semantics e f' g) (p # z) ⟹ semantics e f' g r"
    "all (semantics e f' g) (q # z) ⟹ semantics e f' g r" **for** *f'*
  then show *"semantics e f g r"* **proof** *(simp, metis)* **qed**
next
  fix *f p q z* **assume** *"all (semantics e f g) z"*
    "all (semantics e f' g) z ⟹ semantics e f' g (Con p q)" **for** *f'*
  then show *"semantics e f g p"* *"semantics e f g q"*
  proof *(simp, metis, simp, metis)* **qed**
next
  fix *f p z q c* **assume** *∗: "all (semantics e f g) z"*
    "all (semantics e f' g) z ⟹ semantics e f' g (Exi p)"
    "all (semantics e f' g) (sub 0 (Fun c []) p # z) ⟹ semantics e f' g q"
    "news c (p # q # z)" **for** *f'*
  obtain *x* **where** *"semantics (λn. if n = 0 then x else e (n − 1)) f g p"*
    using *∗(1) ∗(2)* **proof** *force* **qed**
  then have *"semantics (put e 0 x) f g p"* **proof** *simp* **qed**
  then have *"semantics (put e 0 x) (f(c := λw. x)) g p"*
    using *∗(4) allhead allnew map* **proof** *blast* **qed**
  then have *"semantics e (f(c := λw. x)) g (sub 0 (Fun c []) p)"*
    proof *(simp add: substitute)* **qed**
  moreover have *"all (semantics e (f(c := λw. x)) g) z"*
    using *∗(1) ∗(4) alltail allnew allmap* **proof** *blast* **qed**
  ultimately have *"semantics e (f(c := λw. x)) g q"* **using** *∗(3)* **proof** *simp* **qed**
  then show *"semantics e f g q"* **using** *∗(4) allhead alltail allnew map*
  proof *blast* **qed**
next

fix f z t p **assume** *"all (semantics e f g) z"*
  "all (semantics e f' g) z $\implies$ semantics e f' g (sub 0 t p)" **for** f'
then have *"semantics (put e 0 (semantics_term e f t)) f g p"*
proof *(simp add: substitute)* **qed**
then show *"semantics e f g (Exi p)"* **proof** *(simp, metis)* **qed**
next
  fix f z t p **assume** *"all (semantics e f g) z"*
    "all (semantics e f' g) z $\implies$ semantics e f' g (Uni p)" **for** f'
  then show *"semantics e f g (sub 0 t p)"* **proof** *(simp add: substitute)* **qed**
next
  fix f c p z **assume** *: *"all (semantics e f g) z"*
    "all (semantics e f' g) z $\implies$ semantics e f' g (sub 0 (Fun c [])) p)"
    "news c (p # z)" **for** f'
  have *"semantics ($\lambda n.$ **if** n = 0 **then** x **else** e (n − 1)) f g p"* **for** x
  proof −
    have *"all (semantics e (f(c := $\lambda w.$ x)) g) z"*
      using *(1) *(3) alltail allnew allmap **proof** blast **qed**
    then have *"semantics e (f(c := $\lambda w.$ x)) g (sub 0 (Fun c []) p)"*
      using *(2) **proof** simp **qed**
    then have *"semantics ($\lambda n.$ **if** n = 0 **then** x **else** e (n − 1))*
        (f(c := $\lambda w.$ x)) g p"
      proof *(simp add: substitute)* **qed**
    then show *"semantics ($\lambda n.$ **if** n = 0 **then** x **else** e (n − 1)) f g p"*
      using *(3) allhead alltail allnew map **proof** blast **qed**
  qed
  then show *"semantics e f g (Uni p)"* **proof** simp **qed**
qed simp_all

The lemma *soundness'* shows the soundness of the proof system. It is done by rule induction on the rules of the proof system. We have to prove that assuming that the derivations in the premises follow logically, then so does the derivation in the conclusion. For the rules *Boole*, *Dis_E*, *Con_E1*, *Con_E2* and *Uni_E* we state the induction hypothesis, and the assumption that the premises are satisfied. We then do the proof by automation. For *Uni_I*, *Exi_E* and *Exi_I* we write out the proofs explicitly because they are more complicated. We prove the remaining rules sound by automation with the substitution lemma as simplification rule. The keyword **next** is used to separate the different cases.

Let us look at how we proved *Uni_I* sound. The * states our induction hypothesis which states that if our assumptions z are satisfied by any function map then so is

p with a constant *Fun c []* substituted for 0.

$$all \ (semantics \ e \ f' \ g) \ z \implies semantics \ e \ f' \ g \ (sub \ 0 \ (Fun \ c \ []) \ p)$$

We additionally assume that the side condition that c is new to $p\#z$.

$$news \ c \ (p \ \# \ z)$$

Since we want to prove the derivation from z to *Uni p* sound we also assume that the premises z are satisfied by a fixed f and a fixed g.

$$all \ (semantics \ e \ f \ g) \ z$$

We then wish to prove that so is *Uni p*. Since the premises are satisfied by f and since c is new to them they must also be satisfied by $f(c := \lambda w. \ x)$.

$$all \ (semantics \ e \ (f(c := \lambda w. \ x)) \ g) \ z$$

In this step we used the proof method *blast* which is a tableau prover [17]. Then it follows by our induction hypothesis that also p with c substituted for 0 is satisfied.

$$semantics \ e \ (f(c := \lambda w. \ x)) \ g \ (sub \ 0 \ (Fun \ c \ []) \ p)$$

We then use the substitution lemma to add the value of t to the environment instead of doing the substitution.

$$semantics \ (\lambda n. \ \textbf{if} \ n = 0 \ \textbf{then} \ x \ \textbf{else} \ e \ (n - 1)) \ (f(c := \lambda w. \ x)) \ g \ p$$

Since c is new to p we might as well evaluate it in f instead of $f(c := \lambda w. \ x)$ and this concludes the proof.

$$semantics \ (\lambda n. \ \textbf{if} \ n = 0 \ \textbf{then} \ x \ \textbf{else} \ e \ (n - 1)) \ f \ g \ p$$

5.6 A Consistency Corollary to the Soundness Theorem

Soundness is the main theorem about the formalization of the natural deduction proof system. As a corollary we immediately prove the following consistency result about the proof system:

Something, but not everything, can be proved.

In Isabelle we can prove it using the simplifier (*simp*), some simple rules and Isabelle's prover for intuitionistic logic (*iprover*), although a classical prover (say, *metis*) would work too, of course:

corollary *"∃p. OK p []"* *"∃p. ¬ OK p []"*
proof −
  have *"OK (Imp p p) []"* ***for*** *p* ***proof*** *(rule Imp_I, rule Assume, simp)* ***qed***
  then show *"∃p. OK p []"* ***proof*** *iprover* ***qed***
  have *"¬ semantics (e :: nat ⇒ unit) f g Falsity"* ***for*** *e f g* ***proof*** *simp* ***qed***
  then show *"∃p. ¬ OK p []"* ***using*** *soundness* ***proof*** *iprover* ***qed***
qed

Recall that ∃ is the existential quantifier in Isabelle. The symbol ¬ is negation in Isabelle. The first part (∃p. OK p [] **for** p) follows from a simple proof of $p \to p$ (for an arbitrary formula p in first-order logic). The second part (∃p. ¬ OK p []) follows from the proof of soundness and from the fact that the semantics of *Falsity* is always false (for simplicity we consider universes with just one element, provided by the *unit* type).

5.7   Style of the Proof

When you do a proof in Isabelle, you need to choose how close you want the steps of the proof to be to each other. On one hand the proof should be understandable, but on the other hand you do not want the readers to get lost in small details. Larger steps also allow the reader to think for himself instead of having everything spelled out in detail. If a student wants to gain more insight, she can expand it, and let Isabelle check if the details she added were correct. Isabelle also has tools that allow its users to see which steps *simp* used to prove a result.

The notation we chose to use is close to that of programming rather than that of mathematics and set theory. Isabelle, however, also supports a more classical notation. Our motivation for the choice is our students' background from programming, as well as to show that a very well-defined structure lies beneath the logical symbols both at the object and the meta levels.

We use the formal semantics and soundness proof in our teaching. Among other things the students can make calculation using the formal semantics in Isabelle and also make changes to the formal semantics (for example, replacing the if-then-else with logical operators in Isabelle, or adding negation to the logic).

6   Related Work

Formalizations of model theory and proof theory of first-order logic are rare, for example [6, 7, 11, 20, 21].

Throughout the development of NaDeA we have considered some of the natural deduction assistants currently available. Several of the tools available share some common flaws. They can be hard to get started with, or depend on a specific platform. However, there are also many tools that each bring something useful and unique to the table. One of the most prominent is PANDA, described in [13]. PANDA includes a lot of graphical features that make it fast for the experienced user to conduct proofs, and it helps the beginners to tread safely. Another characteristic of PANDA is the possibility to edit proofs partially before combining them into a whole. It definitely serves well to reduce the confusion and complexity involved in conducting large proofs. However, we still believe that the way of presenting the proof can be more explicit. In NaDeA, every detail is clearly stated as part of the proof code. In that sense, the students should become more aware of the side conditions to rules and how they work.

Another tool that deserves mention is ProofWeb [10] which is open source software for teaching natural deduction. It provides interaction between some proof assistants (Coq, Isabelle, Lego) and a web interface. The tool is highly advanced in its features and uses its own syntax. Also, it gives the user the possibility to display the proof in different formats. However, the advanced features come at the cost of being very complex for bachelor students and require that you learn a new syntax. It serves as a great tool for anyone familiar with natural deduction that wants to conduct complex proofs that can be verified by the system. It may, on the other hand, prove less useful for teaching natural deduction to beginners since there is no easy way to get started. In NaDeA, you are free to apply any (applicable) rule to a given formula, and thus, beginners have the freedom to play around with the proof system in a safe way. Furthermore, the formalized soundness result for the proof system of NaDeA makes it relevant for a broader audience, since this gives confidence in that the formulas proved with the tool are actually valid.

7 Further Work

In NaDeA there is support for proofs in propositional logic as well as first-order logic. We would also like to extend to more complex logic languages, the most natural step being higher-order logic. This could be achieved using the CakeML approach [8].

Other branches of logic would also be interesting. Apart from just extending the natural deduction proof system to support other branches of logic, another option is to implement other proof systems as well.

Because the NaDeA tool has a formalization in Isabelle of its proof system, we would like to provide features that allow for a more direct integration with Isabelle.

For instance, we would like to allow for proofs to be exported to a format suitable for Isabelle such that Isabelle could verify the correctness of the proofs. A formal verification of the implementation would require much effort, but perhaps it could be reimplemented on top of Isabelle (although probably not in TypeScript / JavaScript) or using Isabelle's code generation facility.

We would like to extend NaDeA with more features in order to help the user in conducting proofs and in understanding logic. For example, the tool could be extended with step-by-step execution of the auxiliary primitive recursive functions used in the side conditions of the natural deduction rules.

NaDeA has been successfully classroom tested in a regular course with around 70 bachelor students in computer science each year. The students find the formal semantics and the proof of the soundness theorem relevant and instructive. We have extended NaDeA with a so-called ProofJudge system [19] which allows students to submit solutions and get feedback. We are in the process of adding to NaDeA a simple automated theorem prover [20, 21], verified by the Isabelle proof assistant and developed using Isabelle's code generation facility, in order to make it possible to better guide the students if for example sub-proofs are started and there is in fact no possible proof.

References

[1] Mordechai Ben-Ari. Mathematical Logic for Computer Science. Third Edition. Springer 2012.

[2] Tobias Nipkow, Lawrence C. Paulson and Markus Wenzel. Isabelle/HOL - A Proof Assistant for Higher-Order Logic. Lecture Notes in Computer Science 2283, Springer 2002.

[3] Dag Prawitz. Natural Deduction. A Proof-Theoretic Study. Stockholm: Almqvist & Wiksell 1965.

[4] Francis Jeffry Pelletier. A Brief History of Natural Deduction. History and Philosophy of Logic, 1-31, 1999.

[5] Melvin Fitting. First-Order Logic and Automated Theorem Proving. Second Edition Springer 1996.

[6] John Harrison. Formalizing Basic First Order Model Theory. Lecture Notes in Computer Science 1497, 153–170, Springer 1998.

[7] Stefan Berghofer. First-Order Logic According to Fitting. Formal Proof Development. Archive of Formal Proofs 2007.

[8] Ramana Kumar, Rob Arthan, Magnus O. Myreen and Scott Owens. HOL with Definitions: Semantics, Soundness, and a Verified Implementation. Lecture Notes in Computer Science 8558, 308–324, Springer 2014.

[9] Jørgen Villadsen, Anders Schlichtkrull and Andreas Viktor Hess. Meta-Logical Reasoning in Higher-Order Logic. Accepted at 29th International Symposium Logica, Hejnice Monastery, Czech Republic, 15-19 June 2015.

[10] ProofWeb. Online `http://proofweb.cs.ru.nl/login.php` (ProofWeb is both a system for teaching logic and for using proof assistants through the web). Accessed September 2016.

[11] Jasmin Christian Blanchette, Andrei Popescu and Dmitriy Traytel. Unified Classical Logic Completeness - A Coinductive Pearl. Lecture Notes in Computer Science 8562, 46–60, 2014.

[12] Krysia Broda, Jiefei Ma, Gabrielle Sinnadurai and Alexander Summers. Pandora: A Reasoning Toolbox Using Natural Deduction Style. Logic Journal of IGPL, 15(4):293–304, 2007.

[13] Olivier Gasquet, François Schwarzentruber and Martin Strecker. Panda: A Proof Assistant in Natural Deduction for All. A Gentzen Style Proof Assistant for Undergraduate Students. Lecture Notes in Computer Science 6680, 85–92. Springer 2011.

[14] Antonia Huertas. Ten Years of Computer-Based Tutors for Teaching Logic 2000-2010: Lessons learned. Lecture Notes in Computer Science 6680, 131–140, Springer 2011.

[15] Michael Huth and Mark Ryan. Logic in Computer Science: Modelling and Reasoning About Systems. Second Edition. Cambridge University Press 2004.

[16] Jonathan P. Seldin. Normalization and Excluded Middle. I. Studia Logica, 48(2):193–217, 1989.

[17] Jasmin Christian Blanchette, Lukas Bulwahn and Tobias Nipkow. Automatic Proof and Disproof in Isabelle/HOL. Lecture Notes in Computer Science 6989, 12-27, 2011.

[18] Tobias Nipkow and Gerwin Klein. Concrete Semantics with Isabelle/HOL. Springer 2014.

[19] Jørgen Villadsen. ProofJudge: Automated Proof Judging Tool for Learning Mathematical Logic. Exploring Teaching for Active Learning in Engineering Education Conference (ETALEE), Copenhagen, Denmark, 2015.

[20] Jørgen Villadsen, Anders Schlichtkrull and Andreas Halkjær. Code Generation for a Simple First-Order Prover. Isabelle Workshop, Nancy, France, 2016.

[21] Alexander Birch Jensen, Anders Schlichtkrull and Jørgen Villadsen. Verification of an LCF-Style First-Order Prover with Equality. Isabelle Workshop, Nancy, France, 2016.

Received 11 October 2016

LEON HENKIN: A LOGICIAN'S VIEW ON MATHEMATICS EDUCATION

MARÍA MANZANO*
Department of Philosophy, Logic and Aesthetics, University of Salamanca, Salamanca, Spain
mara@usal.es

NITSA MOVSHOVITZ-HADAR
Department of Education in Science and Technology, Technion-Israel Institute of Technology, Haifa, Israel
nitsa@technion.ac.il

DIANE RESEK
Departmet of Mathematics, San Francisco State University, San Francisco, California, USA
resek@sfsu.edu

Abstract

This paper is divided into two parts. In the first one reasons are given for strongly recommending reading some of Henkin's expository papers. The second one describes Leon Henkin's work as a social activist in the field of mathematics education, as he invested a large portion of his career to increase the number of women and underrepresented minorities in the upper echelons of the mathematicians' community.

Keywords: Leon Henkin, completeness, mathematical induction, identity, history of logic, logic education.

*This research has been possible thanks to the research project sustained by Ministerio Economía y Competitividad of Spain with reference FFI2013-47126-P.

1 Some of Henkin's expository papers

1.1 Leon Albert Henkin

Leon Albert Henkin (April 19, 1921, Brooklyn, New York - November 1, 2006, Oakland, California) was a logician at the University of California, Berkeley. His first degree was in mathematics and philosophy from Columbia College, in 1941. He was a doctoral student of Alonzo Church at Princeton University, receiving his Ph.D. in 1947. He became Professor of Mathematics at the University of California, Berkeley, where he had a position from 1953. He received the 1964 Chauvenet Prize for exposition. He was a collaborator of Alfred Tarski, and an ally in promoting logic. In 1990 he was the first recipient of the Mathematical association of America's Gung and Hu Award for Distinguished Service to Mathematics.[1]

His writings became influential from the very start of his career with his doctoral thesis, *The completeness of formal systems*. He then published two papers in the Journal of Symbolic Logic, the first, on completeness for first order logic, in 1949, and the second devoted to completeness in type theory, in 1950.

Henkin was an extraordinary insightful professor of mathematics specializing in logic and interested also in mathematics education. He was blessed with unusual writing and speech capabilities and he devoted considerable effort to writing expository papers. Four groups of them are described below in more details, trying to convince you to read them with your graduate students as a source of mutual inspiration.

1.2 Are Logic and Mathematics Identical?

This is the title of a wonderful expository paper [11], which Leon Henkin published in *Science* in 1962, and subtitled: *An old thesis of Russell's is reexamined in the light of subsequent developments in mathematical logic.*

You may wish to give this paper to your graduate students, not only because the historical view provided is comprehensive and synthetic but also because it shows Henkin's characteristic style; namely, his ability to strongly catch the reader's attention from the beginning.

How does he achieve it? you might wonder. In that particular paper, Henkin tells us that his interest in logic began at the age of 16, *'One day I was browsing in the library and came across a little volume of Bertrand Russell entitled **Mysticism and Logic'**.*[2] In the introduction Henkin cites Russell's radical thesis, that *'mathematics*

---

[1] For more information about Leon Henkin read [22], pages 2-22.

[2] All the quotes in this paragraph come from [11], page 788.

was nothing but logic' together with the companion thesis *'that logic is purely tautological'*, and he describes the strong reaction against his thesis by the academic community: *'Aux armes, citoyens du monde mathématique!'*

Henkin then devotes the first section of the paper to elaborate on the two main ideas that could help explain how Russell arrived at his conclusion. The first one was the lengthy effort to achieve a *'systematic reduction of all concepts of mathematics to a small number of them'*,[3] and the second one was *'the systematic study by mathematical means of the laws of logic which entered into mathematical proofs'*.[4] Henkin relates the work of Frege, Peirce, Boole and Schröder, during the second half of the nineteenth century, to the two efforts mentioned above, and identifies them as the primary *raison d'etre* of *Principia Mathematica*.

In the following section, entitled *From Russell to Gödel*, Henkin explains the introduction of semantic notions by Tarski, as well as the formulation and proof of the completeness theorem for propositional logic by Post and for first-order logic by Gödel. *'This result of Gödel's is among the most basic and useful theorems we have in the whole subject of mathematical logic'*.[5] But, Henkin also explains how, in 1931, the hope of further extension of this kind of completeness was *'dashed by Gödel himself [...] (he) was able to demonstrate that the system of **Principia Mathematica**, taken as a whole was **incomplete**'*.[6] Immediately after, and anticipating what the reader might be thinking, Henkin banished the hope of finding new axioms to repair the incompleteness phenomenon.

In the section entitled *Consistency and the Decision Problem*, Henkin analyzes these important notions and also explains how they are related.

> Indeed, in that same 1931 paper to which I have previously referred, Gödel was able to show that the questions of consistency and completeness were very closely linked to one another. He was able to show that *if* a system such as the *Principia* were truly consistent, then in fact it would not be possible to produce a sound proof of this fact!'.
> (Henkin 1962 [11], p. 791)

In the following section, named *Logic after 1936*, Henkin describes how Alonzo Church proved that no decision procedure is available for first-order logic, and he devotes the rest of the paper to set theory, recursive functions, and algebraic logic. The results he cites takes us only to the 1960's. Of course, much has happened in

---

[3]See [11] page 788.
[4]See [11] page 789.
[5]See [11] page 790.
[6]See [11] page 790.

the field since that time. However, Henkin's description of the field between 1936 and 1962 is quite interesting.

The paper ends in a section where Henkin analyzes *Russell's Thesis in Perspective*. He concludes that if one considers that set theory is a part of logic, the basic concepts of mathematics can "be expressed in terms of logic." In this respect, the *'adult'* Henkin agrees with Russell. He concludes, however:

> The fact that *certain* concepts are selected for investigation, from among all logically possible notions definable in set theory, is of the essence. A true understanding of mathematics must involve an explanation of which set-theory notions have "mathematical content," and this question is manifestly not reducible to a problem of logic, however, broadly conceived.
>
> Logic, rather than being all of mathematics, seems to be but one branch. But it is a vigorous and growing branch, and there is reason to hope that it may in time provide an element of unity to oppose the fragmentation which seems to beset contemporary mathematics —and indeed every branch of scholarship.
>
> *(Henkin 1962 [11], pp. 793-794)*

Henkin was awarded the *Chauvenet Prize* in 1964 for this paper. The prize is described as a *Mathematical Association of America* award to the author of an outstanding expository article on a mathematical topic by a member of the Association.

1.2.1 Bertrand Russell's request

In April 1, 1963, Henkin received a rather unusual letter from Bertrand Russell. In it, Russell thanked Henkin for *'your letter of March 26 and for the very interesting paper which you enclosed.'* Right at the beginning Russell declared:

> It is fifty years since I worked seriously at mathematical logic and almost the only work that I have read since that date is Gödel's. I realized, of course, that Gödel's work is of fundamental importance, but I was puzzled by it. It made me glad that I was no longer working at mathematical logic. If a given set of axioms leads to a contradiction, it is clear that at least one of the axioms is false. Does this apply to schoolboys' arithmetic, and if so, can we believe anything that we were taught in youth? Are we to think that 2+2 is not 4, but 4.001? Obviously, this is not what is intended.
>
> *(Reproduced in: Grattan-Guiness [6], p. 592)*

Russell then went on explaining[7] his *'state of mind'* while Whitehead and he were doing the Principia *'What I was attempting to prove was, not the truth of the propositions demonstrated, but their deducibility from the axioms. And, apart from proofs, what struck us as important was the definitions'* and he added: *'Both Whitehead and I were disappointed that the **Principia** was almost wholly considered in connection with the question whether mathematics is logic.'*

Russell ended the letter: *'If you can spare the time, I should like to know, roughly, how, in your opinion, ordinary mathematics —or, indeed, any deductive system— is affected by Gödel's work.'*

According to Annellis:[8] *'Henkin replied to Russell at length with an explanation of Gödel's incompleteness results, in a letter of July 1963, specifically explaining that Gödel's showed, not the inconsistency, but the incompleteness of the [**Principia**] system.'*

1.3   On Mathematical Induction

On Mathematical Induction is the title of another expository paper [9], which Henkin published in 1960. In a personal communication to María Manzano, Henkin wrote: *'[...] my little paper on induction models from 1960, which has always been my favorite among my expository papers.'* In this paper Henkin examined the crucial concept of *definition by mathematical induction* in the framework of Peano's Arithmetic. To reach that objective, the relationship between the induction axiom and recursive definitions is studied in depth.

In many ways, it is the best paper on logic to offer students as a first reading of a "real-life" article. In what follows we will try to support this judgement. Lets us quote some paragraphs from Henkin's introduction:

> According to modern standards of logical rigor, each branch of pure mathematics must be founded in one of two ways: either its basic concepts must be *defined* in terms of the concepts of some prior branch of mathematics, in which case its theorems are deduced from those of the prior branch of mathematics with the aid of these definitions, or else its basic concepts are taken as *undefined* and its theorems are deduced from a set of axioms involving these undefined terms.
>
> The natural numbers, 0, 1, 2, 3,..., are among those mathematical entities about which we learn at the earliest age, and our knowledge of these numbers and their properties is largely of an intuitive character.

---

[7]The quotes in this paragraph come from [6], pp. 592-593.

[8]See [3], p. 89, footnote 3.

Nevertheless, if we wish to establish a precise mathematical theory of these numbers, we cannot rely on unformulated intuition as the basis of the theory but must found the theory in one of the two ways mentioned above. Actually, both ways are possible. Starting with pure logic and the most elementary portions of the theory of sets as prior mathematical sciences, the German mathematician Frege showed how the basic notions of the theory of numbers can be defined in such a way as to permit a full development of this theory. On the other hand the Italian mathematician Peano, taking *natural number*, *zero* and *successor* as primitive undefined concepts, gave a system of axioms involving these terms which were equally adequate to allow a full development of the theory of natural numbers. In the present paper we shall examine the concept of *definition by mathematical induction* within the framework of Peano's ideas.
    (Henkin 1960 [9], p. 323)

In the first section of *On Mathematical Induction*, entitled *Models and the axioms of Peano*, Henkin not only introduces Peano's axioms[9] and the use of the word *model*, but also what he terms *induction models*, namely, models of the induction axiom P3.

   P3. *If G is any subset of N such that (a) $0 \in G$, and (b) whenever $x \in G$ then also $Sx \in G$, then $G = N$.*

Henkin gives some examples of induction models appart from the obvious Peano models. It seems that Henkin's introduction of induction models in this paper had a pedagogical motivation, as we will see below.

In the following section, entitled *Operations defined by mathematical induction*, Henkin recalls that although the axioms for the theory of natural numbers are very important, the most interesting theorems of the theory did not stem from them alone because in most of the theorems, operations of addition, multiplication, etc. are used. As a first example Henkin poses the definition of addition by the usual equations:

1.1. $x + 0 = x$

---

[9]There is no universal agreement about whether to include zero in the set of natural numbers. Some authors begin the natural numbers with 0, corresponding to the non-negative integers $0, 1, 2, 3, \ldots$, (as Henkin did) whereas others start with 1, corresponding to the positive integers $1, 2, 3, \ldots$. because they find it quite odd to include 0 in the natural numbers as history showed how unnatural the concept of zero as a number was to mankind.

1.2. $x + Sy = S(x + y)$

He then poses the question: *'But in what sense do the equations constitute a definition of addition? In particular, does the definition hold only for natural numbers, or for arbitrary Peano models as well?'*

He then considers a more general problem, that of the introduction of new operations in a Peano model.

> The introduction of an operation by means of the pair of equations 1.1. and 1.2. is an example of what is called *definition by mathematical induction*. To describe the concept in general terms we must consider a Peano model $\langle N, 0, S \rangle$ and in addition a second model $\langle N_1, 0_1, S_1 \rangle$ which, however, is not required to be a Peano model (or even an induction model). Being given these two models we say that the pair of equations
> 2.1. $h(0) = 0_1$
> 2.2. $h(Sy) = S_1(hy)$
> defines (by mathematical induction) a function h: a function which maps N into N_1 and satisfies 2.1. and 2.2. for all $y \in N$.
> *(Henkin 1960 [9], p. 325)*

However, this definition must be justified by a theorem in which the existence of a unique operation that will satisfy the previous equations must be established. Henkin proceeds by presenting an argument to prove the existence of such a function h, and he adds his critical judgment of it being a poor one, as follows:

> Clearly (the argument goes), h is defined for 0, since $h0 = 0_1$ by 2.1. Furthermore, if h is defined for an element y of N then h is also defined for Sy since $h(Sy) = S_1(hy)$ by 2.2. Thus if we let G be the set of all those $y \in N$ for which h is defined, we see that (a) $0 \in G$, and (b) whenever $y \in G$ then also $Sy \in G$. Applying Axiom P3 we conclude that $G = N$. Thus h is defined for all $y \in N$.
>
> At first sight this argument may seem convincing, but a moment's reflection will suffice to raise doubts. For in this argument we refer to a certain function h. But what is h? Apparently it is a function which satisfies 2.1. and 2.2. Recall, however, that the argument is designed to establish the existence of such a function; clearly, then, it is incorrect to assume in the course of the argument that we have such a function.
> *(Henkin 1960 [9], p. 327)*

In 1982 Leon Henkin was an honorary guest of the Department of Logic, History and Philosophy of Science of the University of Barcelona,[10] and he gave two talks; one on *mathematical induction* the other on *cylindric algebras*. These algebras were designed as an algebraic approach to first-order logic, much as Boolean algebras provide an algebraic approach to propositional logic. The second talk was especially interesting because Henkin described to the audience something that would never appear in a formal article: his motivation. In particular, induction models were originaly defined in that paper to show *'that the **axiom** of mathematical induction does not itself justity **definitions** by mathematical induction'.*

Henkin told the audience how he was trying to convince a colleague that the above argument was completely wrong, even though at first sight it might seem convincing. Henkin made him see that in the proof only the third axiom was used and that, if correct, the same reasoning could be used not only for models that satisfy all Peano axioms, but also for those that satisfy only the induction one. Henkin proved that in induction models not all recursive operations are definable. For example, exponentiation fails.

Henkin begins the section entitled *The relation between Peano models and induction models* with the question: *'Why is it that the operations of addition and multiplication exist in every induction model, while the existence of an exponential operation can be guaranteed only for Peano models? To answer this, we must first understand the relation which holds between Peano models and more general induction models.'*[11]

The definitive answer comes in the form of Theorem IV from which categoricity of Peano's arithmetic easily follows.

> Theorem IV. Let $\mathfrak{N} = \langle N, 0, S \rangle$ be a Peano model and $\mathfrak{N}_1 = \langle N_1, 0_1, S_1 \rangle$ an arbitrary model. A necessary and sufficient condition that $\mathfrak{N}_1$ be a homomorphic image of $\mathfrak{N}$ is that $\mathfrak{N}_1$ be an induction model.
>
> [...]
>
> Theorem V Any two Peano models are isomorphic.
>
> *(Henkin 1960 [9], p. 333)*

Induction models turn out to have a fairly simple mathematical structure: there are standard ones —that is, isomorphic to natural numbers— but also non-standard ones. The latter also have a simple structure: either they are cyclic, in particular $\mathbb{Z}$

---

[10]A review was published in [19] by María Manzano. During the academic year 1977-1978 she was a Postdoctoral Fulbright student in The Group in Logic and the Methodology of Science at the University of California, Berkeley. Leon Henkin acted as her advisor.

[11]See [9], page 333.

modulo n, or they are what Henkin called "spoons" in his Barcelona presentation; the name came from their special structure, as they have a handle followed by a cycle. The reason is that the induction axiom is never fulfilled alone, since it requires Peano's first or second axiom. This does not mean that Peano's axioms are redundant, as it is well known that they are formally independent; *i.e.*, each one is independent of the other two.

In the last section, *A characterization of Peano models*, Henkin proves a *'characteristic for Peano models: these are the only models in which all definitions by mathematical induction are justified.'*

1.4 Identity as a Logical Primitive

This is the title of an expository paper Henkin wrote in the seventies. In this paper the crucial role played by the identity relation is analyzed in depth. At the very beginning he declares: *'By the relation of identity we mean that binary relation which holds between any object and itself and which fails to hold between any two distinct objects'.*[12]

As logicians we often pose the following questions: *Why do we take identity as a logical primitive concept in first order logic? Is there a formula φ defining it?*

Henkin explains why the answer to the second question is negative, even in the best scenario where we only have a finite set of non-logical predicate constants. In this situation we can express that x and y cannot be distinguished in our formal language by defining a binary relation that obeys the usual rules for equality.

> For instance, if the only non-logical constants of a given theory are a binary predicate symbol G and a unary predicate symbol U, then the relation E is defined by specifying that Exy holds if an only if, the formula
>
> $$\forall z(Gxz \equiv Gyz) \wedge \forall z(Gzx \equiv Gzy) \wedge (Ux \equiv Uy)$$
>
> holds.
> *(Henkin 1975 [16], p. 31)*

However, the definition is not semantically adequate as there are models where the relation defined by this formula is not identity. The formula is the nearest we can come up with in first order logic to formalize *Leibniz's principle of indiscernibles* saying that two objects are identical when there is no property able to distinguish

---

[12]See Henkin [16], p. 31.

them. *'On the other hand, higher-order systems are well known to admit such a definition. E.g. we can define $x = y$ by the formula $\forall G(Gx \equiv Gy)$'*.[13] This formula can be used to define identity for individuals as the relation defined by it is "genuine" identity in any standard second order structure.

Due to the central role the notion of identity plays in logic, you can either be interested in how to define it using other logical concepts, or else, in the opposite direction, namely using identity to define other concepts. In the first case, you investigate what kind of logic is required. In the second one, you become interested in the definition of the other logical concepts (connectives and quantifiers) in terms of the identity relation, using also abstraction. In this paper Henkin is concerned primarily with definitions in the second direction.

The biconditional connective is usually defined using other connectives, but the main question here is the reverse one: how to use identity to obtain the rest. In this particular case, the answer is obvious: *'The identity relation on the set of truth values, T and F, serves as the denotation of the biconditional connective (in extensional logic, to which we restrict ourself)'*.[14] We know that in propositional logic we are not able to define connectives, such as conjunction, the truth table of which shows a value T on an odd number of lines, not even with equality and negation. The easy explanation is given by Henkin:

> If we take any formula built up from propositional variables with biconditional as the sole connective, or for that matter if we allow negation as well as biconditional, its truth table will be found to have the value T on an even number of lines.
> *(Henkin 1975 [16], p. 32)*

Henkin goes on explaining how we can allow quantification over propositional variables of all types (including at least second order propositional variables) and then all connectives are defined with equality and quantifier. This is not the end of the story, as Henkin asks:

> But what about quantifiers —can they in turn be defined in terms of equality? Quine was the first to observe that this is possible in a system where a variable-binding abstractor is present.
> *(Henkin 1975 [16], p. 33)*

We know that the idea of reducing the other logical concepts to identity is an old one which was tackled with some success in 1923 by Tarski [27], who solved the

---

[13]Henkin 1975 [16], p. 32.
[14]Henkin 1975 [16], p. 32.

case for connectors; three years later, Ramsey [26] raised the whole subject; it was Quine [25] who introduced quantifiers in 1937. It was finally answered in 1963 by Henkin [13], in an article where he developed a system of propositional type theory. His paper was followed by Andrews' improvement [1]: *A reduction of the axioms for the theory of propositional types.*[15] In [16] Henkin presents his definition with some details but, for pedagogical reasons, instead of using lambda notation he uses set abstraction.

> The use of Church's λ-notation for functional abstraction is not as widely known as the notions of set —and relational— abstraction made familiar by Russell and Whitehead through the *Principia*. It has therefore seemed to me worthwhile to sketch below a "translation" of the ideas of my earlier paper (as improved by Andrews), setting now in the context of a formulation of simple type theory based on equality and relational abstraction.
>
> *(Henkin 1975 [16], pp. 33-34)*

In this paper Henkin uses a relational type theory covering all types built up from individual as well as propositional types:

> In addition to variables, the language $\mathcal{L}$ is to be provided with *constants*. For each type symbol a there is a constant symbol $Q_{(aa)}$ used to denote the element of $\mathcal{D}_{(aa)}$ which is the equality relation over $\mathcal{D}_a$.
>
> Finally, the language is to contain three improper symbols, $\{,\}$, and $|$, to be used in forming notations for relational abstracts, as described below.
>
> *(Henkin 1975 [16], p. 34)*

In this formal language, Henkin introduces by *'conventions of abbreviation with respect to formulas of $\mathcal{L}$ '*, the common connectives as well as the universal quantifier:

---

[15] Andrew wrote (in [2], page 69) a very touching story telling about this improvement:

> *'On several occasions, I suggested to Henkin that he simply incorporate my proofs into his paper, but he insisted that I publish a separate paper presenting these proofs, and he wrote a very complimentary letter to Andrzej Mostowski (the editor of Fundamenta Mathematicae) recommending that my paper be published immediately following his own paper'.*

(1) $(A_a \equiv B_a)$ abbreviates $Q_{(aa)}A_aB_a$, where A_a and B_a are any formulas of type a.

(2) $(\forall X_a)A_0$ abbreviates $\{X_a \mid A_0\} \equiv \{X_a \mid X_a \equiv X_a\}$

(3) T_0 abbreviates $(\forall x_0)(x_0 \equiv x_0)$, and F_0 abbreviates $(\forall x_0)x_0$

(4) $\neg_{(0)}$ abbreviates $\{x_0 \mid x_0 \equiv F_0\}$

(5) $\wedge_{(00)}$ abbreviates $\left\{ x_0y_0 \mid \forall G_{(0)}[(G_{(0)}x_0 \equiv G_{(0)}T_0) \equiv x_0] \right\}$

(Henkin 1975 [16], p. 36)

Henkin also introduces a deductive calculus for this language as well as the standard interpretation for it containing the domain $\mathcal{D}_0$ of the two truth values, an arbitrary non-empty domain $\mathcal{D}_1$ for individuals, and the whole collection of relations defined over $\mathcal{D}_{a_1}, ..., \mathcal{D}_{a_n}$ for type $(a_1...a_n)$; namely, $\mathcal{D}_{(a_1...a_n)} = \wp(\mathcal{D}_{a_1} \times ... \times \mathcal{D}_{a_n})$.

Due to the incompleteness phenomenon, to be commented in our next section, Henkin also introduces *frames* and *generalized models* for this higher-order language. To finish the paper, Henkin mentions another interpretation taking $\mathcal{D}_0$ as a many valued domain. This interpretation providing a completeness result in the Boolean sense.

> Conversely, we obtain a new completeness result: Whenever $\models_{\mathcal{B}} A_0$ for every complete Boolean algebra $\mathcal{B}$, then also $\vdash A_0$. We may characterize this result by saying that our deductive system is complete in the *Boolean sense*.
>
> [...]
>
> Proofs for the results on $\mathcal{B}$-models will be given in a forthcoming paper.
>
> *(Henkin 1975 [16], p. 43 and footnote 17)*

Unfortunately, it seems that Henkin never published that paper.

1.5 Completeness

This section is a reduced version of the chapter *Henkin on Completeness* [21] that was published in 2014 in the book *The Life and Work of Leon Henkin: Essays on His Contributions* [20]. We highly recommend that book to all people interested in the influential figure of Leon Henkin.

If you take a look at the list of documents Leon Henkin left us, the first published paper, *The completeness of first order logic* [7], corresponds to his well known result, while the last, *The discovery of my completeness proof* [18], is extremely interesting as autobiography, thus ending his career with a sort of fascinating loop.

It seems to the authors of this paper that reading the last paper is a must. Why?
As you know, Leon Henkin authored an important collection of papers, some of them
extremely exciting, as his proof of the completeness theorem both for the theory of
types and for first-order logic. He did so by means of an innovative and highly
versatile method, which was later to be used in many other logics, even in those
known as non-classical.

1.5.1 Henkin's work on completeness

The completeness theorem establishes the correspondence between deductive calculus and semantics. Gödel had solved it positively for first-order logic and negatively
for any logical system able to contain arithmetic. The lambda calculus for the theory
of types [5], with the usual semantics over a standard hierarchy of types, was able
to express arithmetic and hence could only be incomplete. Henkin showed [8] that if
the formulas were interpreted in a less rigid way, accepting other hierarchies of types
that did not necessarily have to contain all the functions but at least the definable
ones, it is easily seen that all consequences of a set of hypotheses are provable in the
calculus. The valid formulas with this new semantics, called *general semantics*, are
reduced to coincide with those generated by the rules of calculus.

In his 1996 paper, we learn about the process of reaching this discovery, which
observed facts he was trying to explain, and why he ended up discovering things that
were not originally the target of his enquiries. Thus, in this case we do not have to
engage in risky hypotheses or explain his ideas on the mere basis of the later, cold
elaboration in scientific articles. It is well known that the *logic of discovery* differs
from what is adopted on organizing the final exposition of our research through their
different propositions, lemmas, theorems and corollaries.

We also learn that the publication order of his completeness results ([7] and [8])
is the reverse order of his discovery of the proofs. The completeness for first-order
logic was accomplished when he realized he could modify the proof obtained for type
theory in an appropriate way. We consider this to be of great significance, because
the effort of abstraction needed for the first proof (that of type theory) provided a
broad perspective that allowed him to see beyond some prejudices and to make the
decisive changes needed to reach his second proof. In [21] you can find a detailed
commentary of Henkin's contribution to the resolution and understanding of the
completeness phenomena.

1.5.2 Henkin's expository papers on completeness

In 1967 Henkin published two very relevant expository papers on the subject we are considering here, *Truth and Provability* [14] and *Completeness* [15], which were published in *Philosophy of Science Today* [24].

Truth and Provability In less than 10 pages, Henkin gives a very intuitive introduction to the concept of truth and its counterpart, that of provability, in the same spirit of Tarski's expository paper *Truth and Proof* [28]. The latter was published in *Scientific American* two years after Henkin's contribution. This is not so surprising as by then Henkin had been working with Tarski in Berkeley for about 15 years and the theory of truth was Tarski's contribution.

The main topics Henkin introduces (or at least touches upon) are very relevant. They include the *use/mention* distinction, the desire for *languages with infinite sentences* and the need for a *recursive definition of truth*, the *language/metalanguage* distinction, the need to avoid reflexive paradoxes, the concept of *denotation* for terms, and the interpretation of *quantified formulas*. He also explains what an *axiomatic theory* is and how it works in harmony with a *deductive calculus*. Properties such as *decidability* and *completeness/incompleteness of a theory* are mentioned at the end. The way these concepts are introduced is admirable, with such élan, and the chain Henkin establishes, which shows how each concept is needed to support the next one.

Completeness In this short expository paper Henkin explores the complex landscape of the notions of completeness. He introduces the notion of logical completeness —both weak and strong— as an extension of the notion already introduced of "completeness of an axiomatic theory". This presentation differs notably from the standard way these notions are introduced today where, usually, the completeness of the logic precedes the notion of completeness of a theory and, often, to avoid misunderstandings, both concepts are separated as much as possible, as if relating them were some sort of terrible mistake or even anathema. Gödel's incompleteness theorem is presented, as well as its negative impact on the search for a complete calculus for higher-order logic. The paper ends by introducing his own completeness result for higher-order logic with general semantics. The utilitarian way Henkin uses to justify his general models as a way of sorting the provable sentences from the unprovable ones in the class of valid sentences (in standard models) is remarkable. We highly recommend that students study this paper thoroughly.

2 Leon Henkin's Roles of Action and Thought in Mathematics Education

Henkin was often described as a social activist. He invested a significant portion of his career to increase the number of women and of underrepresented minorities in the upper echelons of the mathematicians' community. He was also very aware that we are beings immersed in the crucible of history from which we find it hard to escape, an awareness he brought to the very beginning of his interesting article about the teaching of mathematics:

> Waves of history wash over our nation, stirring up our society and our institutions. Soon we see changes in the way that all of us do things, including our mathematics and our teaching. These changes form themselves into rivulets and streams that merge at various angles with those arising in parts of our society quite different from education, mathematics, or science. Rivers are formed, contributing powerful currents that will produce future waves of history.
>
> The Great Depression and World War II formed the background of my years of study; the Cold War and the Civil Rights Movement were the backdrop against which I began my career as a research mathematician, and later began to involve myself with mathematics education.
> *(Henkin 1995 [17], p. 3)*

In this paper he gave both a short outline of the variety of educational programs he created and/or participated in, and interesting details about some of them.

2.1 NSF Summer Institutes.

The **N**ational **S**cience **F**oundation is an independent federal agency created by the United States' Congress in 1950. As you can read in their web page,[16] its aim was *'to promote the progress of science; to advance the national health, prosperity, and welfare; to secure the national defense...'*. Nowadays, NSF is the *'only federal agency whose mission includes support for all fields of fundamental science and engineering, except for medical sciences.'* NSF's Strategic Plan includes Investing in Science, Engineering, and Education for the Nation's Future.

NSF began summer institutes to improve the teaching of mathematics in high school in 1955. In [17] Henkin related this initiative to historical facts: *'The launching of Sputnik demonstrated superiority in space travel, and our country responded*

---

[16]See *http://www.nsf.gov/about/*

in a variety of ways to improve capacity for scientific and technical developments.[17] The launching of Sputnik in 1957 did drive the U.S. Congress to greatly increase the funding for the summer institutes. The increased funding allowed NSF to expand the institutes to college teachers, as well as those from high schools.

Henkin worked with high school teachers in the summer of 1957, shortly before the launching of Sputnik. In the two succeeding summers, he worked with college teachers. His course was aimed at showing the teachers how the facts and procedures they taught their students could be derived in a deductive system. Also he wanted them to use mathematical language with greater precision. He was surprised at the wide variety of mathematical and pedagogical points of view that he found among both the high school and college teachers. Having studied at a top high school and college, he wrote *'my previous experience as a student in high school and college math classes gave me a very narrow view of the nature of instruction around the country.*[18]

The subject of his courses was the axiomatic foundation of number systems. One of his aims was to help instructors understand *'the idea of a proof'* because he believed that it could help instructors in the effort of finding proofs their own, in a much better way than the mere understanding of the steps that constitutes a proof. The success of the courses led him to write a text [12], with some colleagues from the institute. His aim was to allow teachers who could not attend his summer sessions to learn on their own about the deductive system behind their teaching. He found that this text was not sufficient for teachers to learn about proving on their own. In hindsight, 20 years later, he saw that students need experiences in coming up with proofs on their own. He states *'having experienced a search is a big help in understanding someone else's proof.*[19]

2.2   MAA Math Films.

The **M**athematical **A**ssociation of **A**merica was established one century ago, in 1915. As you can read in their web page: [20] *'Over our first century, MAA has certainly grown, but continues to maintain our leadership in all aspects of the undergraduate program in mathematics'.* Long before internet resources became available, there was still interest in using technology in education. The technology available at the time was television and movies. The MAA decided to try its hand at making movies, actually filmed lectures. As Henkin said: *'Sensing a potential infusion of*

---

[17]See [17], p. 4.
[18]See [17], p. 5.
[19]See [17], p. 6.
[20]See *http://www.maa.org/about-maa/maa-history*

technology into mathematics instruction, MAA set up a committee to make a few experimental films. [...] the committee approached me in 1959-60 with a request to make a filmed lecture on mathematical induction which could be shown at the high-school-senior/college-freshman level. I readily agreed:[21] The film was part of the Mathematics Today series, and was shown on public television in New York City and in high schools.[22]

In [17] Henkin described the preparation of the film, both from a technical point of view and from a methodological and pedagogical perspective. He attributed the lack of understanding of the induction principle at the undergraduate level to the formulation as a mathematical principle about sentences.

> These formulations are the source of most student confusion on this subject. It is extremely difficult to be mathematically correct in talking about sentences and their relation to numbers, and it is simply not a suitable subject for beginning students. The principle of induction is, of course, a statement about sets of numbers satisfying two simple conditions; formulated in this way, it is a fine vehicle for giving students practice in forming and using sets of numbers to show that all natural numbers possess various properties.
> *(Henkin [17], p. 6.)*

Henkin began his process of constructing his video-lectures, by first trying out a lecture on high school students. The assigned producer, Larry Dawson, had no background in mathematics, but accompanied Henkin at his trial lecture. Henkin was astonished at the amount of insight this non-mathematician brought to clarifying his lecture. Leon proceeded to try the revised lecture on college students. In all, he tried out the lecture seven times, each time finding ways to improve it. He cites in [17] that this should be a lesson to all teachers who want to improve their teaching: namely, practice and refining are important. Finally, after Henkin's last trial, the producer said it was all quite clear, except for the question: *why should a student care about mathematical induction? Why was it worth learning?*

Leon pondered about it and came up with the following answer:

> [...] mathematical induction is really of great importance to engineering, for it enters into the proofs of a great many of the most fundamental

---

[21] See [17], p. 6.

[22] The Fourth International Congress on Tools for Teaching Logic (http://ttl2015.irisa.fr/) was celebrated in Rennes in 2015. An important part of the *Special Session on Leon Henkin* was to show the movie.

theorems in the branch of mathematics we call analysis —and these theorems are used over and over by engineers.

And yet, to me, the true significance of mathematical induction does not lie in its importance for practical applications. Rather I see it as a creation of man's intellect which symbolizes his ability to transcend the confines of his environment.

After all, wherever we go, wherever we look in our universe, we see only finite sets: the eggs in a market, the people in a room, the leaves in a forest, the stars in a galaxy —all these are finite. But somehow man has been able to send his imagination soaring beyond anything he has ever seen, to create the concept of an infinite set. And mathematical induction is his most basic tool of discovery in this abstract and distant realm.

(Henkin 1960 [10], pp. 10-11)

2.3   Activities To Broaden Opportunity.

"The sixties" is the term used to describe the counterculture and revolution movement that took place in several places in the U.S.A. and Europe. Berkeley students were taking energic actions against segregation in southeastern U.S.A. as well as against military actions in Vietnam. In [17] Henkin said

In the midst of this turmoil I joined in forming two committees at Berkeley which enlarged the opportunity of minority ethnic groups for studying mathematics and related subjects.

[...] We noted that while there was a substantial black population in Berkeley and the surrounding Bay Area, our own university student body was almost "lily white" and the plan to undertake action through the Senate was initiated.

(Henkin 1995 [17], p. 9)

2.3.1   Special Opportunity Scholarship Program

In 1964, Leon Henkin and Jerzy Neyman, a world-famous Polish-American statistician from Berkeley University, started a program at Berkeley to increase the number of minority students entering college from Bay Area high schools. Henkin told us that the initiative came after Neyman's participation in *'the MAA's Visiting Mathematician Program in Fall 1963. He lectured in southern states where, by law, whites and blacks studied in separate colleges. Upon returning to Berkeley he told some*

of his friends that "first-rate students were being given a third-rate education"[23]
Henkin and Neyman undertook actions through the Academic Senate at Berkeley, and in 1964 the Senate established a committee with the desired effect. The committee recruited promising students and offered them summer programs to study mathematics and English. If they persisted in the program, they were offered special scholarships.

The program began with a summer program for 9th and 10th grade students from African American and Hispanic communities whose teachers felt they held academic promise. The students were taught English in the mornings by special high school teachers. In the afternoons, math and English graduate students arranged experiences in various departments on the campus for the students to interest them in attending college.

The program evolved into an academic program for minority students which provided academic, as well as, financial support. The program in mathematics also recruited women students and later began recruiting women and minorities into the graduate mathematics program. With Henkin's backing, the mathematics department allotted 10% of the available graduate admission places and 10% of the teaching assistantships to students from the Special Opportunity Program. Henkin was chair of the committee in charge of the program from its inception until a year before his death in 2006.

2.3.2 Project SEED

In 1964, Henkin heard a talk by a Berkeley High School teacher, Bill Johntz. At that time, Berkeley was racially segregated in the elementary schools, since the African American and Caucasian students lived in different sides of the town. Later, Berkeley was one of the first American communities to institute bussing so that schools could be less racially segregated. Johntz noted that virtually no African American students ever passed elementary algebra, the first course in the high school. He had the idea of teaching them algebra while they were in elementary school. David Page, a mathematician from the University of Illinois and Robert Davis, a university mathematician who at that time was at the University of Syracuse had both begun teaching algebra to elementary school students.

Johntz attended a summer program given by Robert Davis, where Davis used a Socratic group-discovery method to teach the students. Johntz began using the method to teach African American 5th graders algebra and invited Henkin to see him in action. Henkin saw that Johntz was able to raise great enthusiasm in the

---

[23]In [17], p. 9.

class. Significantly, students enjoyed and actively engaged in the process of learning, and they became integrally involved in their own education.

Henkin found some university funding and began recruiting graduate students to teach using Johntz' methods in the African American elementary schools in Berkeley. The method worked well, and more teachers were needed than were available among Berkeley's graduate students. Engineers and others were loaned by their companies. The program was called Project SEED — Special Elementary Education for the Disadvantaged. This program is still alive, as you can see in their web page. [24]

Henkin and Johntz went to the California legislature and got funding for a system wide program throughout the campuses of the University of California to pay mathematics graduate students to teach minority elementary students throughout the state using the methods from Project SEED. The new program was called Community Teaching Fellowship Program (CTFP). This program encouraged many mathematics graduate students to become involved in pre-college teaching.

2.4   Educating Teachers and Students

2.4.1   Cambridge Conference and its effect on Henkin

In 1963 some leading mathematicians and mathematics educators met in Cambridge Massachusetts to discuss pre-college mathematics curriculum reform. They tried to imagine what a radically different curriculum for mathematics could be in the distant future. They were thinking about the year 2000. They devised the outlines of a curriculum by which high school graduates would have learned the first three years of what was then and what is now an undergraduate mathematics major. They drew up some broad pedagogical guidelines but gave little thought to the education of teachers who would be teaching the reform material.

Henkin was not present at that 1963 conference, but he was invited to its successor in 1966, which focused on the training of teachers. In his paper [17], he states: '*Well. I participated in the Second Conference, but can find no account in my memory or my bookshelves of what proposals we constructed*.'[25] However, when he returned to Berkeley after the conference, he was moved to involve himself in the University mathematics department course for prospective elementary school teachers, so this was the beginning of his involvement in mathematics education.

Previously, the course for elementary school teachers had been taught by lecturers from the College of Education, but Henkin elicited the help of his prominent colleague, J.L.Kelley, and they took over the teaching of the course. Henkin began

---

[24]See http://projectseed.org/
[25]In [17], p. 13.

writing a text for the course, which in his paper *The Roles of Action and of Thought in Mathematics Education* he claims:[26] *'would have resembled several others produced by mathematicians at that time in which elements of modern algebra were presented as a way of 'explaining' the familiar algorithms of school arithmetic.'* After trying out a preliminary version with students, Henkin was led to discontinue the project. He later reflected upon the project, after he had begun to work with teachers: *'I came to believe that the emotional responses of the teachers to mathematics was of more importance to the learning process of the students than the teacher's ability to relate the algorithms of arithmetic to the axioms of ring theory.'*

2.4.2 Film Project and its Effect

Henkin's next encounter with teacher education came in 1968 when he was asked by the National Council of Teachers of Mathematics to serve on a committee to help produce a series of films for teachers on the rational numbers. Along with 12 films to educate teachers about the rational numbers, the project produced a text for teachers. Henkin wrote the second chapter of the book: *Fractions and Rational Numbers*. The chapter was a rather formal presentation of the subject which Henkin later felt contained some *'good ideas'* but *'may be difficult for teachers to follow.'*

In fact, ten years later in 1978, he hand-drafted an abstract and an introduction entitled: *"Logical and pedagogical foundations for the theory of non-negative rational numbers"* which he sent to his former student, Nitsa Movshovitz-Hadar, then a mathematics education lecturer at Technion, Israel Institute of Technology, as a platform for collaborative work. In 1979, Henkin spent a part of his Sabbatical year at the Technion. During that period he collaborated with Movshovitz-Hadar to complete his formal presentation and to combine it with pedagogical methods. Specifically, Henkin proposed five different models for 'founding' the notion of a positive rational number and each model was supposed to be paired with a pedagogical method for presenting fractions to children. Further work on this marvelous idea was carried out in 1995 as the mathematics educator, Pearla Nesher joined in and more work was done in later years. This paper was never completed, and Henkin's unfinished work on it can be found in [23].

2.4.3 1969-1973 National Science Foundation Project for Berkeley Elementary Schools

Henkin's participation in the conference in Cambridge and the film project brought him into contact with the well-known mathematician and mathematics educator,

---

[26]The quotes in this paragraph come from [17], p. 13.

Robert B. Davis. As was mentioned above, the town of Berkeley was moving from segregated schools to integrated ones. The move was challenging to teachers as the African American students lacked self-confidence and there was an academic achievement gap between the different races. Henkin and Davis applied to the National Science Foundation for a grant to help some elementary school teachers become mathematics specialists in their schools and to find ways of making all students more successful in mathematics. When the project was funded, Diane Resek, a student of Henkin's at the time, was selected as the coordinator. The project consisted of a summer program and then meeting throughout the academic year. Davis was in charge of the summer program and Henkin of the academic component. Henkin worked closely with Resek and occasionally conducted sessions himself for the teachers. The teachers were charmed by Henkin and seemed to feel privileged by the attention and respect he paid them.

2.4.4 Bay Area Math Project and the American Mathematics Program

In 1982, Judy Kysh, now a professor of mathematics and mathematics education at San Francisco State University, started a summer program at the University of California, Davis, to educate pre-college teachers so they could provide a more engaging and coherent math curriculum. The State of California instituted the California Mathematics Project based on the Davis program and others. Its mission was to improve the pre-college teaching of mathematics especially for students from underrepresented populations. A second purpose was to use both pre-college teachers and university faculty to work together to educate other pre-college teachers mathematically and pedagogically. The teachers and faculty were to have equal status, using their complementary expertise to better educate others.

In 1983, Henkin spearheaded a team from University of California, Berkeley to establish a site for the project in the Berkeley Area. It was called the Bay Area Mathematics Program and is still in operation at this time. Henkin worked with Lyle Fisher, a high school teacher. At times, Henkin sat in as a participant when high school teachers led hands-on workshops on topics such as tessellations. At the other times he invited visiting mathematicians such as Henry Pollack from Columbia University to address the teachers on mathematics topics such as applications of mathematics.

A *few* years later Henkin worked with Kysh and others to establish the American Mathematics Project through the Mathematics Association of America. The project was funded for three years and was based on the same principle of combining the complementary expertise of mathematicians and mathematics educators. Six teams of pairs, consisting of a faculty member and a precollege teacher, from around the

United States, took part in the program. Henkin was committed to valuing mathematical and pedagogical expertise equally in educating teachers. He wanted to get more mathematicians involved in mathematics education.

2.4.5 Project 2061

The American Association for the Advancement of Science (AAAS) began an ambitious program in 1986, the last time Halley's Comet passed the earth. By 2061, the next time the comet would come by, the AAAS hoped that all Americans would be truly literate in mathematics, science, and technology. To begin the project, they chose prominent mathematicians and scientists to delineate what they felt a truly scientific literate person would need to know in 2061. Leon Henkin and David Blackwell, a prominent statistician from the University of California, Berkeley were asked to lead a panel of renowned university mathematicians and statisticians to draw up a list of the important mathematical ideas a high school graduate should possess in 2061 in order to be scientifically literate. Because of Henkin's respect for the knowledge of pre-college mathematics teachers and other mathematics educators, he added several consultants to advise the panel.

The report of the committee [4], which was written by Henkin and Blackwell, contains a well-reasoned response to the question: What are "the important ideas of mathematics that everyone should know and understand"? It is also, as are most documents penned by Henkin, a joy to read.

The document lays out in sections: the processes of mathematics, the subject areas of mathematics, mathematics in science and technology, mathematics and language, emotions and mathematics, and, finally, concluding remarks. The first of the concluding remarks states: *'Our over arching theme is that mathematics is a part of human experience; it emerges from every-day experience and can be reflected back on it.'* The remarks go on to tie in the other sections of the report. Two important remarks are:

> Mathematical learning should be integrated with play. Many mathematical ideas should emerge from a variety of constructions and other projects having physical, chemical, biological elements, as well as from games possessing economic and strategic characteristics.
>
> [...]
>
> Children should be helped to develop intuitive ideas about 'how things work' in various realms of experience. They should learn how to 'translate' these intuitions into hypotheses about mathematical models of real world phenomena, and they should get used to adjusting intuitions and models to fit within experience.

(In Blackwell and Henkin [4], p. 60)[27]

2.5   Open Sesame: The Lawrence Hall Of Science

In 1968 the University of California, Berkeley instituted a science museum on the Berkeley campus, called the Lawrence Hall of Science. It was created in honor of the 1939 Alfred Nobel prize winner Ernest Orlando Lawrence. As you can read in their web page: [28]

'We have been providing parents, kids, and educators with opportunities to engage with science since 1968.'

A physics professor, Alan Portis, was made the first director. Portis wanted the Hall to become more than a museum; he conceived of a center for science and math education at all levels. He enlisted the participation of scientists, mathematicians and science educators to form a graduate program of research. The participants created a PhD program named, SESAME, Search for Excellence in Science and Math Education.

Henkin and John Kelley were the two mathematicians on the faculty of the program. Although the physicists and biologists began working with graduate students, Henkin and Kelley stayed away from that aspect of the program. They did institute a math education study group with graduate students in math and math education, who met regularly to inform themselves about new ways of teaching math. Henkin also became one of three co-directors of the Hall and through that experience gained familiarity with hands-on exhibits to teach mathematics.

In Spring 1972, Nitsa Movshovitz-Hadar wrote him asking to work with him on a PhD in mathematics education through SESAME. She was then a student from the Technion in Israel and now is a retired professor of math education from the Technion, a former director of the Israel National Museum of Science, Technology, and Space, and a contributor to this paper. Henkin relates his reaction to her request in [17]

> She seemed well qualified, had strong recommendations, and I felt she ought to be admitted. But how could I take on the responsibility of directing her research, when I had never myself pursued research in math education (despite the many projects in which I had worked)? I felt very insecure and uneasy. Finally, however, I said to myself. "If not me, with whom should such a student work?" I had no good answer for this question, so reluctantly said OK.

---

[27]Quote appears on the final draft sent to Nitsa Movshovitz-Hadar on Jan 26, 1988 from Henkin.
[28]See http://www.lawrencehallofscience.org/about

(Henkin 1995 [17], p. 16)

Henkin was a dedicated thesis advisor. He is remembered running a Xerox machine to get copies of a student's thesis to the rest of the student's committee while the student was writing the last pages. He encouraged his students to write more practical theses that could affect classrooms and learning directly. Elizabeth Stage, then a lecturer in the SESAME program and now the Director of the Lawrence Hall of Science, remembers that this perspective often put him at odds with other SESAME faculty, who took more ideological and theoretical points of view. At one defense of thesis meeting, a physicist who was committed to behaviorism, asked Henkin's PhD student why the material the student had devised was asking elementary students to figure out the answers. "Why not tell them how to look at it." Henkin defended his student by saying the point was whether students could learn without being told.

Henkin *'fathered'* four other PhD students in SESAME after working with Movshovitz-Hadar. He reflected on the beginning of his work with SESAME students in [17]:

> The path of my work in mathematics education is about to change its character, and its role within my professional work. The emphasis up to this point has been on activities stimulated by external events, which generated thought to accomplish defined goals. Beyond this point we shall see an internal growth and development of ideas relating to mathematics education such as we are used to experiencing when we do mathematics; and new ideas will follow from these thoughts, rather than lead them.
>
> *(Henkin 1995 [17], p. 16)*

Henkin promised at the end of [17] to write another paper describing his new phase. Unfortunately, Henkin never produced that second paper. He did not leave any finished papers in mathematics education. We have lost a lot by not hearing how his ideas on mathematics education finally unfolded.

References

[1] P. Andrews. A reduction of the axioms for the theory of propositional types. *Fundam. Math.* 52:345–350, 1963.

[2] P. Andrews. A Bit of History Related to Logic Based on Equality. In: *The Life and Work of Leon Henkin: Essays on His Contributions.* pp. 67-71. Studies in Universal Logic. Springer International Publishing. Switzerland, 2014.

[3] I. Anellis.: Review of Bertrand Russell, Towards the "Principles of Mathematics", 1900-02 , edited by Gregory H. Moore, and Bertrand Russell, Foundations of Logic, 1903-05 edited by Alasdair Urquhart with the assistance of Albert C. Lewis. *Mod. Log.* 8(3-4):57–93, 2001. http://projecteuclid.org/euclid.rml/1081173771.

[4] D. Blackwell and L. Henkin. Mathematics: report of the Project 2061 Phase I Mathematics Panel. Published by the American Association for the Advancement of Science. Project 2061. 1988.

[5] A. Church. A formulation of the simple theory of types. *J. Symb. Log.* 5:56-68, 1940.

[6] I. Grattan-Guinness. *The Search for Mathematical Roots, 1870-1940. Logics, Set Theories, and the Foundations of Mathematics from Cantor Through Russell to Gödel.* Princeton University Press. Princeton, 2000.

[7] L. Henkin. The completeness of the first-order functional calculus. *J. Symb. Log.* 14(3):159–166, 1949.

[8] L. Henkin. Completeness in the theory of types. *J. Symb. Log.* 15(2):81–91, 1950.

[9] L. Henkin. On mathematical induction. *The American Mathematical Monthly* 67(4):323–338, 1960.

[10] L. Henkin. Mathematical induction. Mathematical Association of American Film Manual no. 1, 21 pp. Manual to accompanymotion picture "Mathematical induction", 1960.

[11] L. Henkin. Are Logic and Mathematics Identical? *Science* 138:788–794, 1962.

[12] L. Henkin et als. *Retracing Elementary Mathematics*, Macmillan, New York, 1962

[13] L. Henkin. A theory of propositional types, *Fundam. Math.* 52:323–344, 1963.

[14] L. Henkin. Truth and provability. In: Morgenbesser, S. (ed.). *Philosophy of Science today*, New York: Basic Books, 1967

[15] L. Henkin. Completeness. In: Morgenbesser, S. (ed.). *Philosophy of Science today*, New York: Basic Books, 1967.

[16] L. Henkin. Identity as a logical primitive, *Philosophia* 5:31–45, 1975.

[17] L. Henkin. The Roles of Action and of Thought in Mathematics Education–One Mathematician's Passage. In: Fisher, N. D., Keynes, H. B., Wagreich, Ph. D. (eds.). *Changing the Culture: Mathematics Education in the Research Community.* CBMS Issues in Mathematics Education, vol. 5, pp. 3-16. American Mathematical Society in cooperation with Mathematical Association of America, Providence, 1995.

[18] L. Henkin. The discovery of my completeness proofs. *The Bulletin of Symbolic Logic* 2(2):127–158, 1996.

[19] M. Manzano. Conferències del professor Leon Henkin de la Universitat de California (Berkeley). In: *Ciència. Revista catalana de Ciència y tecnologia.* 24:10-11, 1983

[20] M. Manzano et als (eds.).: *The Life and Work of Leon Henkin: Essays on His Contributions.* Studies in Universal Logic. Springer International Publishing. Switzerland (2014)

[21] M. Manzano. Henkin on Completeness. In: Manzano, M. et als (eds.). *The Life and Work of Leon Henkin: Essays on His Contributions.* pp. 149-175. Studies in Universal Logic. Springer International Publishing. Switzerland, 2014.

[22] M. Manzano and E. Alonso. Leon Henkin. In: Manzano, M. et als (eds.). *The Life and Work of Leon Henkin: Essays on His Contributions.* pp. 2-22. Studies in Universal Logic. Springer International Publishing. Switzerland, 2014.

[23] N. Movshovitz-Hadar. Pairing Logical and Pedagogical Foundations for the Theory of Positive Rational Numbers-Henkin's Unfinished Work. In: Manzano, M. et als (eds.). *The Life and Work of Leon Henkin: Essays on His Contributions.* pp. 149-175. Studies in Universal Logic. Springer International Publishing. Switzerland, 2014.

[24] S. Morgenbesser. (ed.). *Philosophy of Science today,* New York: Basic Books. 1967.

[25] W. Quine. Logic based on inclusion and abstraction. *J. Symb. Log..* 2:145–152, 1937

[26] F. P. Ramsey.: The foundations of mathematics, in: *Proceedings L. M. S.* (2)25:338-384, 1926.

[27] A. Tarski.: Sur le terme primitif de la logistique in: *Fundam Math.* 4:196–200, 1923.

[28] A. Tarski.: Truth and Proof. *Scientific American,* June 1969, 63–70, 75–77, 1969.

          Received 11 October 2016

Teaching Modal Logic from The Linear Algebraic Viewpoint

Ryo Hatano, Katsuhiko Sano and Satoshi Tojo
Japan Advanced Institute of Science and Technology, Japan
{r-hatano, v-sano, tojo}@jaist.ac.jp

Abstract

This paper proposes a linear algebraic approach to teach modal logic to students who might not be familiar with first-order logic. Our approach is based on Fitting's linear algebraic reformulation of Kripke semantics of modal logic. A key idea of his reformulation is to represent an accessibility relation R by a square matrix and a valuation $V(p)$ of an atomic variable p by a column vector. Then, we may calculate the truth set of $\Diamond p$ as the multiplication of the square matrix R for the accessibility relation and the column vector for p. Hence, we can regard such matrix calculation as an extended version of truth table calculation. We discuss how our reformulation is useful to teach modal logic to our target students before teaching first-order logic. In addition, we present our supporting software to avoid involved calculations on matrices and explain how we can use it for educational purposes.

Keywords: Modal logic, Linear Algebra, Quantification, First-order logic.

1 Introduction and Motivation

1.1 Linear Algebraic View for Kripke Semantics of Modal Logic

In order to teach modal logics to students effectively, we propose to use Fitting's linear algebraic approach. Under the approach, we can teach many elementary topics of Kripke semantics for modal logics by calculations over Boolean matrices. In this paper, our target students are those who have prior knowledge of both linear algebra and propositional logic. In general, they are first or second year undergraduate students at the departments of computer science, electrical engineering, and physics. They might not be familiar with first-order logic.

The authors thank the anonymous referees for their helpful comments.

Modal logics are often taught to students as one of advanced topics after propositional logic and first-order logic. This is because Kripke semantics of modal logics relies on knowledge of quantification and binary relation of first-order logic and model theory. In particular, the notions of existential quantification and universal quantification are used to define the semantics of $\diamond$ and $\square$ operators over the Kripke model, respectively. Moreover, quantifications are also used to define frame conditions of frame properties, e.g., for all w, wRw reflexivity holds (wRv stands for 'there is a link from w to v'). Therefore, the ordinary target students of a course on modal logic are assumed to belong to the departments of philosophy, computer science, and mathematics. Such students might be familiar with first-order logic and its model-theoretic semantics. In particular, they might have already learned the syntactic notions of quantification and binary relation of first-order logic and the model-theoretic explanation of them.

So far, many topics of Kripke semantics are taught to students by the model-theoretic approach. Some topics of modal logic are easier to understand than those of first-order logic, since we can explain Kripke semantics on a graphical representation of a Kripke model (see the left side of Figure 1). For example, we can calculate the truth value of a formula over a graphical representation of a Kripke model visually. Under this approach, however, there are some topics which might confuse our target students. For example, such topics include the truth of $\square p$ at a 'dead-end' world where we cannot access any world, and the verification of the Euclideanness property (wRv and wRu jointly imply vRu, for all w, v, u). In order to show the truth of $\square p$ at the dead-end world, we need to teach students when the implication is vacuously true. This might be unnatural for some students. In addition, the verification of Euclideanness of a given frame might be difficult for some students, since they need to check whether the frame satisfies the condition of Euclideanness very carefully where v and u are possibly the same. If the cardinality of the domain of the model is larger, such checking might be more involved.

In order to teach the above topics of modal logics to students effectively, we propose to use Fitting's linear algebraic reformulation of Kripke semantics. A key idea of his reformulation is to represent an accessibility relation R by a Boolean square matrix and a valuation $V(p)$ of an atomic variable p by a Boolean column vector, provided the cardinality of the domain is finite (see the right side of Figure 1).[1] As a result, we may compute the truth set of a formula by calculations over Boolean matrices. For example, the truth set of $\diamond p$ is calculated by the multiplication of the square matrix of R and the vector of $V(p)$. Moreover, we may also verify the

---

[1] We note that this assumption is often justified because most of the well-known modal logics, for example **T**, **S4** and **S5**, enjoy the finite model property, i.e., φ is the theorem of a modal logic Λ iff φ is valid for all *finite* models for the logic Λ.

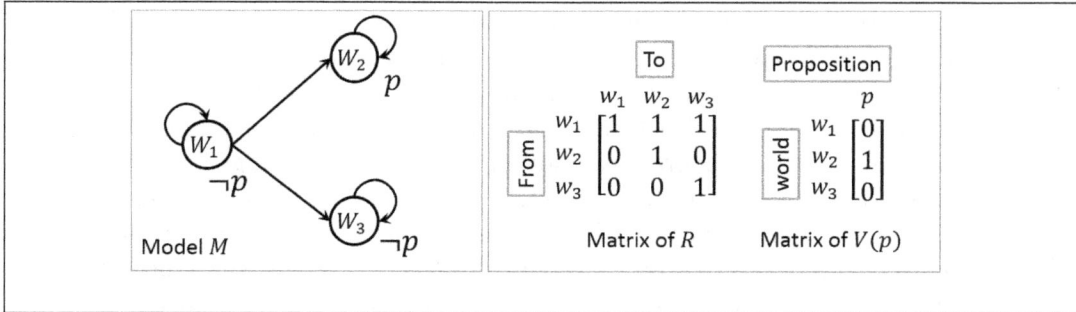

Figure 1: Kripke model and its Boolean matrix representation

frame property of a given frame by the calculation of matrices. Since these calculations are based on the truth-table calculation of propositional logic, we can regard the calculations as an extended version of truth-table calculation. As a result, we may replace required prior knowledge of the quantifications and binary relation of first-order logic by the truth-table calculation of propositional logic and elementary calculations of Boolean matrices of linear algebra. Moreover, this extension allows us to calculate some restricted forms of quantifications (in Kripke semantics) *without* bound variables of first-order logic.

We claim that our linear algebraic approach is helpful to students who have prior knowledge of both linear algebra and propositional logic. Using our approach, they can learn modal logics based on their acquired knowledge without prior knowledge of first-order logic. Moreover, ordinary target students, who have already learned first-order logic, may deepen their understanding of the subject from a different perspective. In order to obtain feedbacks from students, we have held a small seminar to teach elementary topics of modal logic using our approach. The details will be explained in Section 4.5.

1.2 Related Studies

Our key idea comes from Fitting's linear algebraic reformulation of Kripke semantics for (multi-) modal logics [4]. He explained that a machinery of matrices over Boolean algebras is appropriate to investigate multi-modal semantics. In his paper, a Kripke frame is regarded as a directed graph and represented by a Boolean matrix, and a valuation of each atomic variable is represented by a Boolean vector. Then, a linear algebraic reformulation of Kripke semantics is defined by Boolean operations on matrices and vectors. This idea is also extended to define relations between frames, however, we employed the above reformulation only.

The other idea of our linear algebraic approach comes from the following previous works. Liau [9] introduced Boolean matrix operations for multiple agents' belief reasoning, revision, and fusion. Based on the matrix representation of belief states, he proposed a belief logic and its algebraic semantics. Similarly, Fusaoka et al. [5] introduced real-valued matrix operation for qualitative belief change in a multi-agent system. Based on the above studies, Tojo [14] proposed notions of Boolean matrices and vectors for the simultaneous informing action with communication channels. He showed that the notions of matrices can represent a public announcement [11] and a consecutive message passing. It can be regarded as an application of linear algebraic approach to multi-agent communication. Then, Hatano et al. extended Tojo's idea to provide rigorous definitions in [6]. They proposed a decidable and semantically complete multi-agent doxastic logic with communication channels and its dynamic extensions with two informing action operators. With the help of van Benthem and Liu's idea of *relation changer* [16], their dynamic operators can be regarded as program terms in propositional dynamic logic. Afterward, they provided a linear algebraic reformulation of the proposed semantics. In addition, a supporting software based on the above idea is also provided. The present paper expands Hatano et al.'s work into the educational content. In connection with spatial logics and linear algebra, we refer to a survey by van Benthem and Bezhanishvili [15]. In the survey, they mentioned connections between modal logic and linear algebra over vector spaces $\mathbb{R}^n$. Different from our approach, they did not provide Kripke semantics using Boolean matrices.

Finally, we also refer to relevant study to ours in terms of relational algebra by Berghammer and Schmidt [2]. They proposed a relational algebraic approach to investigate finite models of non-classical logics such as multi-modal logics with common knowledge operators and computational tree logic. They interpret the logics in relation algebra with transitive closures whose representation is based on Boolean matrices. They also present applications of their tool RELVIEW based on the above idea for finite model checking tasks, e.g., the muddy children puzzles. Their study seems very close to ours, although their approach is different from ours since they simply used Boolean matrices as an input and output interface to their computation system and did not compute the matrices directly.[2] For example, they defined a 'composition' of two Boolean square matrices which corresponds to two accessibility relations by a componentwise relational composition, although we define it by a multiplication of Boolean matrices. Another difference is that in this paper, we have a list of the matrix representation of frame properties and two types of

---

[2]Relations can be represented by Boolean matrices [12], although ordinary works of relational formalization do not employ such representation, e.g., a study of relational formalization of non-classical logics by Orlowska [10].

correspondence between modal axioms and their matrix representations.

1.3 Outline of This Paper

The rest of this paper is organized as follows. In Section 2, we recall basic definitions and notions of modal logic and Boolean matrices. Section 3 explains a matrix representation of Kripke semantics and its relevant properties, and also connect our argument to the concept of quantification in first-order logic. In Section 4, we explain how we can use our approach for educational purposes. We explain which teaching topics of modal logics can be taught to students using our approach and why the approach is helpful for educational purposes. Moreover, we present our supporting software to avoid involved calculations on matrices. In addition, we introduce our teaching experiment and summarize feedbacks from students. Finally, in Section 5, we summarize our contribution and conclude with further remarks.

2 Preliminaries

2.1 Modal logic

First, we recall the ordinary (propositional) modal logic. A *modal language* L is composed of the following vocabulary: A finite set $\mathsf{Prop} = \{p, q, r, \dots\}$ of propositional letters; Boolean connectives $\neg, \vee$; diamond operator $\Diamond$. A set of formulas of L is inductively defined as follows:

$$\varphi ::= p \mid \neg\varphi \mid (\varphi \vee \varphi) \mid \Diamond\varphi$$

where $p \in \mathsf{Prop}$. We will omit the parentheses whenever convenient. A formula $\Diamond\varphi$ stands for 'it might be the case that φ'. We introduce the dual operator of $\Diamond$ by $\Box\varphi := \neg\Diamond\neg\varphi$, whose reading is 'it must be the case that φ', and the Boolean connectives $\wedge, \to$ as usual abbreviations.

Then, we introduce Kripke semantics for the syntax L. A *Kripke model* $\mathfrak{M}$ is a tuple (W, R, V) where W is a non-empty set of *possible worlds*, called *domain*, $R \subseteq W \times W$ is an *accessibility relation*, and $V : \mathsf{Prop} \to \mathcal{P}(W)$ is a *valuation function*. A *frame* is the result of dropping a valuation function from a model, i.e., (W, R) (denoted by, for instance, $\mathfrak{F}$). We denote $(w, v) \in R$ also by wRv. We use both notations depending on the context. Given any model $\mathfrak{M} = (W, R, V)$ and any possible world $w \in W$, the *satisfaction relation* $\mathfrak{M}, w \models \varphi$ is defined inductively as follows:

$$
\begin{aligned}
\mathfrak{M}, w &\models p & \text{iff} \quad & w \in V(p), \\
\mathfrak{M}, w &\models \neg\varphi & \text{iff} \quad & \mathfrak{M}, w \not\models \varphi, \\
\mathfrak{M}, w &\models \varphi \vee \psi & \text{iff} \quad & \mathfrak{M}, w \models \varphi \text{ or } \mathfrak{M}, w \models \psi, \\
\mathfrak{M}, w &\models \Diamond\varphi & \text{iff} \quad & \mathfrak{M}, v \models \varphi \text{ for some } v \text{ with } wRv.
\end{aligned}
$$

A *truth set* $[\![\varphi]\!]_{\mathfrak{M}}$ is defined by $[\![\varphi]\!]_{\mathfrak{M}} = \{\, w \in W \mid \mathfrak{M}, w \models \varphi \,\}$. Then, we can obtain the following:

$$
\begin{aligned}
{[\![p]\!]_{\mathfrak{M}}} &= V(p), \\
{[\![\neg\varphi]\!]_{\mathfrak{M}}} &= W \setminus [\![\varphi]\!], \\
{[\![\varphi \vee \psi]\!]_{\mathfrak{M}}} &= [\![\varphi]\!]_{\mathfrak{M}} \cup [\![\psi]\!]_{\mathfrak{M}}, \\
{[\![\Diamond\varphi]\!]_{\mathfrak{M}}} &= \{\, w \mid wRv \text{ and } v \in [\![\varphi]\!]_{\mathfrak{M}} \text{ for some } v \,\}.
\end{aligned}
$$

We say that φ is *valid* on a model $\mathfrak{M}$ if $\mathfrak{M}, w \models \varphi$ for all worlds $w \in W$. We also say that φ is valid on a frame $\mathfrak{F}$ if $(\mathfrak{F}, V) \models \varphi$ for all valuation V on $\mathfrak{F}$. In connection with the validity of a formula and the truth set of a formula, the following proposition holds:

Proposition 1. Given a model $\mathfrak{M}$ and a formula φ,

$$\varphi \leftrightarrow \psi \text{ is valid on } \mathfrak{M} \text{ iff } [\![\varphi]\!]_{\mathfrak{M}} = [\![\psi]\!]_{\mathfrak{M}}.$$

Example 2. Recall Figure 1 of Section 1, i.e., we define the model $\mathfrak{M}$ by:

$$
\begin{aligned}
W &= \{\, w_1, w_2, w_3 \,\}, \\
R &= \{(w_1, w_1), (w_1, w_2), (w_1, w_3), (w_2, w_2), (w_3, w_3)\}, \\
V(p) &= \{\, w_2 \,\}.
\end{aligned}
$$

By definition, it is clear that $\Diamond p$ is true at w_1 and w_2, i.e., $\mathfrak{M}, w_1 \models \Diamond p$ and $\mathfrak{M}, w_2 \models \Diamond p$, respectively. In order to compare the ordinary model-theoretic approach with our linear algebraic approach later, we explain the proof of $\mathfrak{M}, w_1 \models \Diamond p$ by the model-theoretic approach. Let us rewrite our goal $\mathfrak{M}, w_1 \models \Diamond p$ by definition as:

For some $v \in W(w_1 R v$ and $\mathfrak{M}, v \models p)$.

By the clause for propositional variable, this is equivalent to:

For some $v \in W(w_1 R v$ and $v \in V(p))$.

By definition, $V(p) = \{\, w_2 \,\}$. In order to obtain our goal, it suffices to know if $(w_1, w_2) \in R$. Since this trivially holds, we conclude $\mathfrak{M}, w_1 \models p$, as required.

In Example 3 in Section 3.1, we will see that we can show $\mathfrak{M}, w_1 \models \Diamond p$ without the notions of the existential quantification. In other words, the above model-theoretic proof can be represented by a simple calculation over Boolean matrices.

2.2 Boolean Matrix

Mathematical operations and some properties of Boolean matrices are slightly different from real-valued matrices. For example, the inverse operation of multiplication seems not well-defined,[3] and the addition of the same matrices satisfies idempotence, i.e., the resultant matrix of the addition is equal to the original one.

Throughout this paper, we use the symbol M, to denote a *Boolean matrix*, i.e., each element of the matrix belongs to the set $\{0, 1\}$. We use the symbol M as a superscript ^{M} with a symbol or expressions (e.g., X^M and $(X + Y)^M$) to denote a matrix representation of them. If the representing matrix is clear from the context, we omit 'M' from such 'X^M' and just write 'X'. Moreover, $M(m \times n)$ means the set of all (Boolean) $m \times n$ matrices, where m and n are the numbers of rows and columns, respectively. In the usual sense, $1 \times n$ and $m \times 1$ matrices are called Boolean row vector and column vector, respectively. Let M be an $m \times n$ matrix, $1 \leq i \leq m$ and $1 \leq j \leq n$. $M(i, j)$ denotes the element in the i-th row and j-th column entry. Moreover, $\boldsymbol{E}$, $\boldsymbol{0}$ and $\boldsymbol{1}$ denote the *unit square matrix* ($\boldsymbol{E}(i, j) = 1$ if $i = j$; 0 otherwise), *complete matrix* ($\boldsymbol{1}(i, j) = 1$ for all i and j), and *zero matrix* ($\boldsymbol{0}(i, j) = 0$ for all i and j), respectively.[4]

The Boolean operations of addition '+', multiplication '$\cdot$' and complement '$^-$' for the element of Boolean matrices correspond to the logical operations of '$\vee$', '$\wedge$' and '$\neg$', respectively. These operations are also defined to the level of matrices. Let $M, M_1, M_2 \in M(m \times n)$. For all i and j, the complement $\overline{M}$ and the addition $M_1 + M_2$ are defined by:

$$
\begin{aligned}
\overline{M}(i, j) &:= \overline{M(i, j)}, \\
(M_1 + M_2)(i, j) &:= M_1(i, j) + M_2(i, j).
\end{aligned}
$$

Given any $M_1 \in M(m \times l)$ and any $M_2 \in M(l \times n)$, the multiplication $M_1 M_2$ of matrices is defined by:

$$
(M_1 M_2)(i, j) = \sum_{1 \leq k \leq n} (M_1(i, k) \cdot M_2(k, j)).
$$

The transposition $^t M$ is defined as: $^t M(i, j) = M(j, i)$ for all i and j. In the below, we summarize basic properties of addition, multiplication and transposition

---

[3]The inverse operation of the Boolean addition, i.e., subtraction is not well-defined over Boolean values. Consequently, subtraction for a Boolean matrix cannot make sense.

[4]Dimensions of those matrices depend on the context.

of Boolean matrices. For any $M, M_1, M_2 \in M(m \times n)$,

$$
\begin{aligned}
M &= M + M, \\
M &= \boldsymbol{E}M, \\
M &= {}^t({}^t(M)), \\
{}^t(M_1 + M_2) &= {}^tM_1 + {}^tM_2, \\
{}^t(M_1M_2) &= {}^tM_2 {}^tM_1.
\end{aligned}
$$

These facts will be used in Section 3.2 to show some propositions of frame properties in matrix representation. For more general introduction to Boolean matrix theory, see [7].

3 Linear Algebraic Reformulation of Kripke Semantics

3.1 Kripke Semantics in Matrices

In this section, we establish a connection between Kripke semantics and its matrix representation with the help of Fitting's idea [4]. Regarding a possible world as a vertex and a tuple (v, u) in an accessibility relation as a directed edge, a frame (W, R) forms a directed graph. If the set of possible worlds is finite, the graph can be represented by a finite adjacency matrix [5] with boolean values, i.e., *Boolean* matrix. In order to focus our discussion on such matrices, we use the following convention.

Convention 1. In what follows in this paper, we restrict our attention to the finite Kripke models.

Informally, Fitting's idea of reformulation of Kripke semantics can be summarized as follows: An accessibility relation (or frame) forms a directed graph and can be represented by a Boolean matrix. A valuation of a proposition (or a truth set of formula) can also be represented by a Boolean (column) vector. Then, propositional connectives correspond to Boolean operations over Boolean vectors, and $\diamond$ operator corresponds to the multiplication of a Boolean matrix and a vector.

Example 3. Recall a Kripke model $\mathfrak{M} = (W, R, V)$ in Example 2 (see also Figure 1). A Boolean vector which represents a truth set of $\diamond p$ can be obtained by a multiplication of the matrix corresponding to R and the column vector corresponding to

---

[5]Let $V = \{v_1, \ldots, v_n\}$ and $E \subseteq V \times V$, we can form a finite directed graph (V, E). Then the adjacency matrix of E is a $n \times n$ square matrix such that its component $M(i, j) = 1$ if there is an edge from vertex i to vertex j, and 0 otherwise.

$V(p)$:

$$\begin{bmatrix} 1 & 1 & 1 \\ 0 & 1 & 0 \\ 0 & 0 & 1 \end{bmatrix} \begin{bmatrix} 0 \\ 1 \\ 0 \end{bmatrix} = \begin{bmatrix} 1 \\ 1 \\ 0 \end{bmatrix}.$$

The resultant vector exactly corresponds to the truth set $[\![\Diamond p]\!]_{\mathfrak{M}} = \{w_1, w_2\}$ in Example 2.

We also emphasize that the calculation of the semantics can be regarded as an extension of truth-table calculation. A truth value of $\Diamond p$ at w_1 is computed by a multiplication of the row vector corresponding to w_1 row of the square matrix of R and the column vector of $V(p)$:

$$\begin{bmatrix} 1 & 1 & 1 \end{bmatrix} \begin{bmatrix} 0 \\ 1 \\ 0 \end{bmatrix} = 1.$$

The resultant value also corresponds to the result of $\mathfrak{M}, w_1 \models \Diamond p$ in Example 2.

Now, let us introduce our linear algebraic reformulation of Kripke semantics in full detail. Let $\mathfrak{F} = (W, R)$ be a (finite) Kripke frame and suppose that the cardinality of W is m and $W = \{w_1, w_2, \ldots, w_m\}$. A matrix representation of an accessibility relation $R^M \in M(m \times m)$ is defined by

$$R^M(i, j) = \begin{cases} 1 & \text{if } (w_i, w_j) \in R, \\ 0 & \text{if } (w_i, w_j) \notin R. \end{cases}$$

Intuitively, a row of the matrix means 'from' world and a column means 'to' world. In order to obtain a matrix representation of Kripke model, it suffices to consider a valuation function in terms of Boolean matrices. Given a Kripke model $\mathfrak{M} = (\mathfrak{F}, V)$ and an atomic proposition $p \in \mathsf{Prop}$, a matrix representation of $V(p)$ is defined to be a column vector $V(p)^M \in M(m \times 1)$ such that

$$V(p)^M(i) = \begin{cases} 1 & \text{if } w_i \in V(p), \\ 0 & \text{if } w_i \notin V(p). \end{cases}$$

The semantic clauses of each formula φ can be defined by the computation over the column vector(s) $\|\varphi\| \in M(m \times 1)$ inductively as follows:

$$\begin{aligned} \|p\| &:= (V(p))^M, \\ \|\neg\varphi\| &:= \overline{\|\varphi\|}, \\ \|\varphi \vee \varphi\| &:= \|\varphi\| + \|\varphi\|, \\ \|\Diamond\varphi\| &:= R^M\|\varphi\|, \end{aligned}$$

where $p \in$ Prop. Note that we may extend our syntax and semantics to a multi-modal language. Let G be a finite set of indices. For syntax, we use $\diamondsuit_a$ operator in multi-modal language instead of $\diamondsuit$ operator and the other operators are the same as mono-modal language. For semantics, an accessibility relation R is replaced by $(R_a)_{a \in G}$, where $R_a \subseteq W \times W$, and their matrix representation becomes R_a^M. Therefore, $\|\diamondsuit_a \varphi\| := R_a^M \|\varphi\|$. In order to focus our attention on elementary teaching topics in Section 4, we will not explain multi-modal extension in this paper.

Proposition 4. Given any finite Kripke model $\mathfrak{M}$ and any formula φ of L,

$$(\llbracket \varphi \rrbracket \mathfrak{m})^M = \|\varphi\|.$$

Example 5. Let R^M be a 2×2 matrix, $p \in$ Prop and $V(p)^M$ be a 2×1 matrix. Let us write

$$R^M := \begin{bmatrix} r_{11} & r_{12} \\ r_{21} & r_{22} \end{bmatrix} \text{ and } V(p)^M := \begin{bmatrix} x \\ y \end{bmatrix}.$$

Then,

$$
\begin{aligned}
\|\Box p\| &= \|\neg \diamondsuit \neg p\| &&= \overline{\|\diamondsuit \neg p\|} \\
&= \overline{R^M \|\neg p\|} &&= \overline{R^M V(p)^M} \\
&= \overline{\begin{bmatrix} r_{11} & r_{12} \\ r_{21} & r_{22} \end{bmatrix} \begin{bmatrix} x \\ y \end{bmatrix}} &&= \overline{\begin{bmatrix} r_{11} & r_{12} \\ r_{21} & r_{22} \end{bmatrix} \begin{bmatrix} \overline{x} \\ \overline{y} \end{bmatrix}} \\
&= \overline{\begin{bmatrix} r_{11}\overline{x} + r_{12}\overline{y} \\ r_{21}\overline{x} + r_{22}\overline{y} \end{bmatrix}} &&= \begin{bmatrix} (\overline{r_{11}} + x) \cdot (\overline{r_{12}} + y) \\ (\overline{r_{21}} + x) \cdot (\overline{r_{22}} + y) \end{bmatrix}.
\end{aligned}
$$

Thus far, we have explained our linear algebraic reformulation of Kripke semantics. From Examples 3 and 5, we can observe that calculations of the truth set of a formula are based on truth-table calculation of propositional logic. Indeed, if we focus on propositional connectives and restrict the cardinality of the domain to 1, a matrix calculation of the truth set of a formula is essentially the same as the truth-table calculation of the formula. In this sense, we may regard matrix calculations of the truth set of a formula as an extended version of truth-table calculation of propositional logic. In the next section, we will explain that another type of an extended truth-table calculation, namely, the verification of frame properties in terms of Boolean matrices.

3.2 Modal Axioms in Matrices

In order to discuss various frame properties, we now explain that relational union and composition can be defined by matrix addition and multiplication as follows:

given two binary relations $R, S \subseteq W \times W$,

$$(R \cup S)^M = R^M + S^M, \quad (R \circ S)^M = R^M S^M$$

where $R \circ S = \{ (w, v) \mid (w, u) \in R \text{ and } (u, v) \in S \text{ for some } u \in W \}$. From an educational perspective, the reader may wonder if we should teach relation algebra after introducing our linear algebraic approach to modal logic since these operations are originally from Tarski's relation algebra [13]. However, this is not the case. Even if our target students do not have prior knowledge of relational composition and union, we can introduce these operations just as the corresponding operations to matrix addition and multiplication. Therefore, we may even introduce the notions from Tarski's relation algebra based on Boolean matrices.

In addition to the above correspondences, the following equivalences will be helpful in proving correspondence between modal formulas and their matrix representations (e.g., Proposition 11 in Section 3.2).

Proposition 6. Given any $R, S \subseteq W \times W$, $R \subseteq S$ iff $S = R \cup S$ iff $S^M = R^M + S^M$.

Now we can reformulate well-known frame properties in terms of Boolean matrices.

Proposition 7. Every frame property listed in Table 1 can be reformulated in terms of Boolean matrix with elementary matrix calculations as in the Table 1 where $\mathbf{1}$ means a column vector of all 1s.

Name	Frame Condition	Formula		Matrix Reformulation
Reflexive	$\forall w(wRw)$	T	$\Box p \to p$	$R = R + E$
Symmetric	$\forall w, v(wRv \text{ implies } vRw)$	B	$p \to \Box \Diamond p$	$R = {}^t R$ (or $R = {}^t R + R$)
Transitive	$\forall w, v, u(wRv \& vRu \text{ imply } wRu)$	4	$\Box p \to \Box\Box p$	$R = RR + R$
Serial	$\forall w \exists v(wRv)$	D	$\Box p \to \Diamond p$	$R^t R = R^t R + E$ (or $\mathbf{1} = R\mathbf{1}$)[6]
Euclidean	$\forall w, v, u(wRv \& wRu \text{ imply } vRu)$	5	$\Diamond p \to \Box \Diamond p$	$R = {}^t RR + R$

Table 1: Frame properties in this paper

We can verify the five frame properties of a given frame in Table 1 in terms of Boolean matrices.

---

[6]We have found more general result than [2]. Their seriality is described by $L \subseteq R; L$ where R is an accessibility relation, L denotes the universal relation, and $R; L$ denotes the relational composition of R with L. The corresponding matrix representation of their seriality is $L = RL$ where L and R are unit square matrix of a relation. If the cardinality of the domain is 2, we can describe their seriality by $[\mathbf{1}\mathbf{1}] = R[\mathbf{1}\mathbf{1}] = [R\mathbf{1}, R\mathbf{1}]$.

Example 8. Recall a matrix representation of an accessibility relation R in Example 2. For simplicity, we regard R as a square matrix of the relation. We can verify whether R satisfies certain frame properties listed in Table 1 by the computation over matrices. For example, let us check whether R satisfies transitivity. By $R = RR + R$, i.e.,

$$\begin{bmatrix} 1\ 1\ 1 \\ 0\ 1\ 0 \\ 0\ 0\ 1 \end{bmatrix} = \begin{bmatrix} 1\ 1\ 1 \\ 0\ 1\ 0 \\ 0\ 0\ 1 \end{bmatrix} \begin{bmatrix} 1\ 1\ 1 \\ 0\ 1\ 0 \\ 0\ 0\ 1 \end{bmatrix} + \begin{bmatrix} 1\ 1\ 1 \\ 0\ 1\ 0 \\ 0\ 0\ 1 \end{bmatrix}$$

$$= \begin{bmatrix} 1\ 1\ 1 \\ 0\ 1\ 0 \\ 0\ 0\ 1 \end{bmatrix}.$$

Hence, we may conclude that R satisfies transitivity. In a similar manner, we can also check whether R satisfies the other properties. Since $R = R + E$ and $\mathbf{1} = R\mathbf{1}$, R satisfies reflexivity and seriality, respectively. However, by $R \neq {}^t R$, $E \neq {}^t RR + E$ and $R \neq {}^t RR + R$, this R does not satisfy symmetricity, and Euclideanness, respectively. Finally, we may conclude that the model satisfies reflexivity, seriality, and transitivity.

Note that the verification of frame properties in Example 8 can also be regarded as an extended truth-table calculation. This is because each verification of frame property in Example 8 is based on Boolean matrix calculation.

Next, we will establish well-known implications among frame properties in terms of Boolean matrices. In addition, we also show an ordinary proof for comparison.

Proposition 9. Reflexivity and Euclideanness jointly imply symmetry, i.e., $R = R + E$ and $R = {}^t RR + R$ jointly imply $R = {}^t R$.

Proof. Firstly, we observe that if R is reflexive, then the transposition ${}^t R$ is also reflexive, i.e., ${}^t R = {}^t R + E$. Secondly, we rewrite the equation of reflexivity, as follows:

$$\begin{aligned} R &= R + E && \text{(Reflexivity)} \\ &= ({}^t RR + R) + E && \text{(by Euclideanness)} \\ &= ({}^t R + E)R + E && \\ &= {}^t RR + E && \text{(by reflexivity of } {}^t R\text{)}. \end{aligned}$$

Afterward, we get ${}^t R = {}^t RR + E$ by transposing both sides. Since both R and ${}^t R$ are equal to ${}^t RR + E$, we finally obtain $R = {}^t R$. $\qquad\square$

For comparison, we show an ordinary proof with quantifiers as follows. We show that for any $w, v \in W$, wRv implies vRw. Fix any w, v such that wRv. By reflexivity, wRw. By Euclideanness, we obtain vRw from wRv and wRw, as desired.

Proposition 10. Reflexivity and Euclideanness jointly imply transitivity, i.e., $R = R + \boldsymbol{E}$ and $R = {}^{t}RR + R$ jointly imply $R = RR + R$.

Proof. $R = {}^{t}RR + R$ (by Euclideanness) $= RR + R$ (by Proposition 9). $\qquad\square$

We also show an ordinary proof with quantifiers as follows. We show that for any $w, v, u \in W$, wRv and vRu imply wRu. Fix any w, v such that wRv and vRu. By symmetry (Proposition 9), vRw. By Euclideanness, we obtain wRu from vRw and vRu.

In the above ordinary proofs, we had to select the appropriate variables for every application of the conditions of the frame properties. The selection of variables might sometimes be a cause of an error in the proof. On the other hand, we did not need to worry about the selection of variables in the above linear algebraic proofs. Therefore, the linear algebraic proofs seem clear and easy to understand for many our target students.

In order to establish a relationship between modal axioms and frame properties, we follow the idea of Lemmon-Scott axioms or Geach axioms [8, 3]. Namely, we show that there are at least two types of correspondence between modal axioms and their corresponding matrix representations in the Table 1. For simplicity, we are omitting superscript M and regard R as a square matrix.

Proposition 11. Let $n, m, l, k \in \mathbb{N}$ and $p \in \mathsf{Prop}$. For all frames $\mathfrak{F} = (W, R)$,

$$\Diamond^{k}\Box^{l}p \to \Box^{m}\Diamond^{n}p \text{ is valid on } \mathfrak{F} \text{ iff } ({}^{t}R)^{m}R^{k} + R^{n}({}^{t}R)^{l} = R^{n}({}^{t}R)^{l}.$$

Proof. Here R^{-1} denotes the inverse relation of R. We observe that ${}^{t}(R^{M}) = (R^{-1})^{M}$. Fix any frame $\mathfrak{F} = (W, R)$.

$$\Diamond^{k}\Box^{l}p \to \Box^{m}\Diamond^{n}p \text{ is valid on } \mathfrak{F},$$
$$\text{iff } (\Diamond^{-1})^{m}\Diamond^{k}\Box^{l}p \to \Diamond^{n}p \text{ is valid on } \mathfrak{F},$$
$$\text{iff } (\Diamond^{-1})^{m}\Diamond^{k}p \to \Diamond^{n}(\Diamond^{-1})^{l}p \text{ is valid on } \mathfrak{F},$$
$$\text{iff } (R^{-1})^{m} \circ R^{k} \subseteq R^{n} \circ (R^{-1})^{l},$$
$$\text{iff } (R^{-1})^{m} \circ R^{k} \cup R^{n} \circ (R^{-1})^{l} = R^{n} \circ (R^{-1})^{l}.$$

This is equivalent to $({}^{t}R)^{m}R^{k} + R^{n}({}^{t}R)^{l} = R^{n}({}^{t}R)^{l}$. $\qquad\square$

Using the above proposition, we can obtain matrix representations of reflexivity, symmetricity, transitivity, seriality, and Euclideanness in Table 1. In addition, we can obtain another matrix representation of seriality, i.e., $1 = R1$, by the following proposition.

Proposition 12. Let $m \in \mathbb{N}$, $p \in \mathsf{Prop}$ and 1 be a vector of all 1s. For all frames $\mathfrak{F} = (W, R)$,

$$\square^m p \to \Diamond^m p \text{ is valid on } \mathfrak{F} \text{ iff } R^m 1 = 1.$$

Proof. Fix any frame $\mathfrak{F} = (W, R)$.

$$
\begin{array}{lll}
\square^m p \to \Diamond^m p \text{ is valid on } \mathfrak{F} & \text{iff} & \Diamond^m \neg p \vee \Diamond^m p \text{ is valid on } \mathfrak{F}, \\
& \text{iff} & \Diamond^m (\neg p \vee p) \text{ is valid on } \mathfrak{F}, \\
& \text{iff} & \Diamond^m \top \leftrightarrow \top \text{ is valid on } \mathfrak{F}, \\
& \text{iff} & R^m 1 = 1.
\end{array}
$$

$\square$

3.3 Quantifications in Matrices

So far we have explained the matrix reformulation of Kripke semantics in modal logic. Now we begin to extend this approach to capture the behaviors of a universal quantifier $\forall$ and an existential quantifier $\exists$ in first-order logic.

Let us consider the case of the universal (or full) relation, i.e., $R = W \times W$. Then, the semantic clauses of $\square$ and $\Diamond$ becomes:

$$
\begin{array}{lll}
\mathfrak{M}, w \models \square \varphi & \text{iff} & \forall v \in W(wRv \text{ implies } \mathfrak{M}, v \models \varphi), \\
\mathfrak{M}, w \models \Diamond \varphi & \text{iff} & \exists v \in W(wRv \text{ and } \mathfrak{M}, v \models \varphi).
\end{array}
$$

Since R is the universal relation, wRv trivially holds. This implies that these clauses are not restricted by the accessibility relation R. Namely, the clauses can be regarded as:

$$
\begin{array}{lll}
\mathfrak{M}, w \models \square \varphi & \text{iff} & \forall v \in W(\mathfrak{M}, v \models \varphi), \\
\mathfrak{M}, w \models \Diamond \varphi & \text{iff} & \exists v \in W(\mathfrak{M}, v \models \varphi).
\end{array}
$$

In this sense, we may regard the semantic clauses for $\square$ and $\Diamond$ of modal logic as the ones for $\forall$ and $\exists$ of first-order logic, respectively.

We can establish a similar argument in terms of Boolean matrices. In the sense of the matrices, the universal relation R becomes the complete square matrix 1. As a result, computations of $\|\Diamond p\|$ and $\|\square p\|$ come to reflect the above argument.

Example 13. Let $W = \{\, w_1, w_2, w_3 \,\}$, R be the universal relation and $V(p) = \{\, w_2 \,\}$. Since there exists a world w_2 such that p holds, $\Diamond p$ also holds at every world, i.e.,

$$\|\Diamond p\| := R^M \|p\| = R^M V(p)^M = \begin{bmatrix} 1 & 1 & 1 \\ 1 & 1 & 1 \\ 1 & 1 & 1 \end{bmatrix} \begin{bmatrix} 0 \\ 1 \\ 0 \end{bmatrix} = \begin{bmatrix} 1 \\ 1 \\ 1 \end{bmatrix}.$$

However, since p does not hold at w_1 and w_3, $\Box p$ does not hold at every world, i.e.,

$$\|\Box p\| := \overline{R^M \overline{\|p\|}} = \overline{R^M \overline{V(p)^M}} = \overline{\begin{bmatrix} 1 & 1 & 1 \\ 1 & 1 & 1 \\ 1 & 1 & 1 \end{bmatrix} \begin{bmatrix} 1 \\ 0 \\ 1 \end{bmatrix}} = \begin{bmatrix} 0 \\ 0 \\ 0 \end{bmatrix}.$$

Now, let us visualize the distinction between $\exists\forall$ and $\forall\exists$ of first-order logic by matrix representation. Let us consider the situation where $\exists x \forall y R(x, y)$, that is, there is some world x from which all the other worlds are accessible. Then, it means that the x-column is filled with 1s. This observation implies that the property of $\exists y \forall x R(x, y)$ is expressed in terms of Boolean matrix as $(^t\overline{R})\mathbf{1} \neq \mathbf{1}$. In the similar way, in case $\forall x \exists y R(x, y)$, that is, for each row there must be at least one 1 (see Table 2). Thus, the property $\forall x \exists y R(x, y)$ of seriality is expressed in terms of Boolean matrix as: $R\mathbf{1} = \mathbf{1}$.

$\exists y \forall x R(x, y)$	$\forall x \exists y R(x, y)$
$\begin{bmatrix} 1 & 0 & 1 \\ 1 & 0 & 1 \\ 1 & 0 & 1 \end{bmatrix}$	$\begin{bmatrix} 0 & 1 & 0 \\ 0 & 0 & 1 \\ 1 & 0 & 1 \end{bmatrix}$

Table 2: Example of nested quantifications in terms of matrices (in 3×3).

Then, we also establish "$\exists\forall$ implies $\forall\exists$" in term of matrices.

Proposition 14. $(^t\overline{R})\mathbf{1} \neq \mathbf{1}$ implies $R\mathbf{1} = \mathbf{1}$.

Proof. Let R be an $n \times n$ matrix. Let us write

$$R := \begin{bmatrix} r_{11} & \cdots & r_{1n} \\ \vdots & \ddots & \vdots \\ r_{n1} & \cdots & r_{nn} \end{bmatrix}.$$

We show the contrapositive implication and so assume $R\mathbf{1} \neq \mathbf{1}$. Now, the goal is to show $({}^t\overline{R})\mathbf{1} = \mathbf{1}$. The assumption implies that $[r_{i1} \ \cdots \ r_{in}] = {}^t\mathbf{0}$ for some $1 \leq i \leq n$. Fix such i. Then, $[\overline{r_{i1}} \ \cdots \ \overline{r_{in}}] = \overline{[r_{i1} \ \cdots \ r_{in}]} = {}^t\mathbf{1}$. Since

$$ {}^t\overline{R} := \begin{bmatrix} \overline{r_{11}} & \cdots & \overline{r_{n1}} \\ \vdots & \ddots & \vdots \\ \overline{r_{1n}} & \cdots & \overline{r_{nn}} \end{bmatrix} $$

$({}^t\overline{R})\mathbf{1} = \mathbf{1}$ holds by $[\overline{r_{i1}} \ \cdots \ \overline{r_{in}}] = {}^t\mathbf{1}$. $\qquad\square$

4 Linear Algebraic Approach to Teach Modal Logic

In this section, from an educational point of view, we explain which topics of modal logics can be taught to students using our approach and why the approach is helpful for educational purposes. In addition, we introduce our supporting software to avoid involved computations on matrices.

4.1 Teaching Topics on Modal Logic by Linear Algebraic Approach

When we teach modal logics to students, the following topics are often covered:

1. Syntax: how to read modal operators, how to define formulas, the dual definition of modal operators, and the distinction between nested modalities.

2. Kripke semantics: a graphical representation of a Kripke model, the satisfaction relation, how to compute the truth value of a formula at a world, the validity and the satisfiability of a given formula, a counter-model construction, frame properties (reflexivity, symmetricity, transitivity, seriality and Euclideanness), and the correspondence between frame properties and formulas (T, B, D, 4 and 5).

3. Proof theory: Hilbert-style systems, tableau methods, natural deductions, sequent calculi, extensions by modal axioms (T, B, D, 4 and 5).

4. Possible further topics: bisimulation, finite model property, and decidability, complexity, soundness and completeness theorem of modal logics.

Here we focus our attention on elementary topics of items 1-3. For item 1, we should teach how to read modal operators at first. We introduce $\Box$ and $\Diamond$ operators and teach how to read them, i.e., we read $\Box$ as 'it is necessary that' and $\Diamond$ as 'it is possible that.' Then, we also teach the other readings of modal operators. For example, we read $\Box$ operator as 'it is believed that' in doxastic logic, 'it is known that' in epistemic logic, 'it is obligatory that' in deontic logic, 'it will always be the case that' and 'it has always been the case that' in temporal logic. Afterward, we should teach how to write a formula of modal logic by a BNF grammar. If $\Diamond$ operator is contained in the syntax, then we can define the other $\Box$ operator as the dual of the operator $\Diamond$, e.g., $\Box p := \neg\Diamond\neg p$. In addition, we should also teach the distinction between nested modalities, e.g., $\Box\Box p$, $\Box\Diamond p$ and $\Diamond\Box p$. In connection with epistemic logic, we also teach what the positive introspection ($\Box p \to \Box\Box p$) and the negative introspection ($\neg\Box p \to \Box\neg\Box p$) mean. The positive introspection stands for 'If agent knows, he/she knows what he/she knows', and negative introspection stands for 'If agent do not know, he/she knows that he/she do not know.'

For item 2, in order to teach Kripke semantics, we use the model-theoretic approach. A good point of modal logic is that we can calculate the truth value of a formula over a graphical representation of a Kripke model visually. However, such graphical approach sometimes might not work well, e.g., a calculation of the truth value of the formula $\Box p$ at the 'dead-end' world where we cannot access any world. In such a case, we should follow the definition of the satisfaction. We often give a brief introduction to the above five frame properties by the graphical approach intuitively, and then we explain frame conditions of a frame property by the model-theoretic approach rigorously. We also explain well-known implications among the frame properties, e.g., reflexivity and Euclideanness jointly imply transitivity (cf. Section 3.2). Afterward, we should explain correspondence between frame properties and valid formulas, e.g., a frame satisfies Euclideanness if and only if 5 ($\Diamond p \to \Box\Diamond p$) is valid on the frame.

For proof theory of item 3, we should introduce the basic proof system first. For example, the Hilbert-style base system is defined by propositional tautology, the distribution axiom for $\Box$ operator, ($\Box(p \to q) \to (\Box p \to \Box q)$), modus ponens (from φ and $\varphi \to \psi$, we may infer ψ), and the necessitation rule for $\Box$ operator (from φ, we may infer $\Box\varphi$). In addition, we also teach what a proof of the theorem is and what the notion of theorem on the base system is.

Next, we should teach additional well-known modal axioms, i.e., T, B, D, 4, and 5. From these five axioms, we also teach that we can consider 32 different combinations of the axioms, but we can reduce them substantially to 15 combinations. By the 15 combinations of axioms, we can determine 15 different modal logics. For example, we can determine **KT**, **KD**, **K45**, **S4**, and **S5**. Thereafter, we explain some extensions of the base system. For example, if we add the axioms T, B, and 4 to the above Hilbert-style base system, it becomes Hilbert-style system for **S5**.

Our linear algebraic approach can cover some of the above topics. In particular, many topics of Kripke semantics (item 2) and soundness of proof theory (item 3) can be covered. But the topics of syntax (item 1) and proof theory (item 3) cannot be covered. We assume that our target students have prior knowledge of propositional logic and linear algebra, and so our approach might be effective for them. In the following sections, we compare the ordinary model-theoretic approach with our linear algebraic approach.

- In Section 4.3.1, we start with a calculation of the truth value of a formula. We also explain how we can verify the validity of a formula on a model in Section 4.3.2.

- In Section 4.4.1, we explain how to verify frame properties of a frame. In this section, we also mention that we can check whether the frame satisfies reflexivity, seriality, and symmetricity at a glance by the form of a matrix of an accessibility relation.

- In Section 4.4.2, we explain how to show the correspondence between frame properties and valid formulas.

For each section, we also explain how to use our software for educational purposes. Finally, in Section 4.5, we explain our teaching experiment and feedbacks from students.

4.2 Supporting Software to Teach Modal Logic

In Section 3, we regarded a calculation of the truth set of a formula and the verification of frame properties listed in Table 1 as an extended truth-table calculation, i.e., a computation on Boolean matrices. However, similarly to the case of the ordinary model-theoretical approach, we have to give more efforts to compute matrices if the length of a given formula becomes longer or the dimension of a matrix becomes bigger. Such efforts might be required when lecturers provide exercises or prepare teaching materials. If we wish to avoid such efforts on calculations, we had better to implement some supporting tools. In this section, we introduce our supporting software to overcome this issue. We provide an overview and a short instruction on our software in the remaining sections.

We have implemented a supporting software based on our linear algebraic reformulation of Kripke semantics by Java^{TM} 8 programming language and opened for the public.[7] The features of our software can be summarized as follows:

1. We can edit a matrix representation of a Kripke model by a graphical user interface easily.

---

[7]http://cirrus.jaist.ac.jp:8080/soft/bc.

2. A computation program of the truth set of a formula on a model is provided. We can obtain a Boolean vector representation of the truth set of the formula written in TEX style, e.g., 'p ¥land q.'[8] From this vector, we can obtain a truth value of the formula for each world and also verify the validity of the formula on the model.[9]

3. A verification program of frame properties is also provided. We can verify all frame properties listed in Table 1 at once.

4. A visualization program is provided. By the program, and we can obtain a graphical representation of a Kripke model via Graphviz.[10]

The provided programs might be helpful for educational purposes. For example, lecturers can use our software to design exercises and lecture materials. In addition, students can use our software to study modal logics by themselves. Notice that unlike RELVIEW tool [2], if we know how to input a formula and a model into our software, we can work with modal logic by our software without any more preparation. RELVIEW tool is designed to solve computation tasks of relation algebra. In order to work with modal logic by RELVIEW tool, we need to provide definitions of formulas and semantics of modal operators based on relational operations to RELVIEW tool by the internal language of it. For example, we need to define $\Box$ operator by box(S, v) = - (S * -v) where S is a matrix for an accessibility relation, v is a vector for valuation and the operators - and * are relational complementation and composition, respectively. However, in the context of the efficiency of computation tasks, we do not claim any superiority of our program over RELVIEW.

Figure 2 shows a sample of the graphical user interface of our software.[11] The interface is divided into two parts. The left side of the interface is an editor for Kripke model, and the right side is a calculator for the computation tasks that we mentioned in the above list of features.

A Kripke model editor allows us to manage a model easily. The design of the editor reflects our approach; namely, we can input the model into the editor by the matrix representation of the model. A general workflow to input the parameters of the model is described as follows:

---

[8] As a matter of practical convenience, there are insert buttons of a proposition, logical connectives, and modal axioms at the next to the parameter box of the calculator. Hence, we can input the above vocabulary of modal logic written in TEX style into the parameter box easily.

[9] We note that our software was originally introduced in [6] to support computation tasks for dynamic logic of multi-agent communication.

[10] http://www.graphviz.org/

[11] Displaying parameters are corresponding to the formula and the model in Example 2.

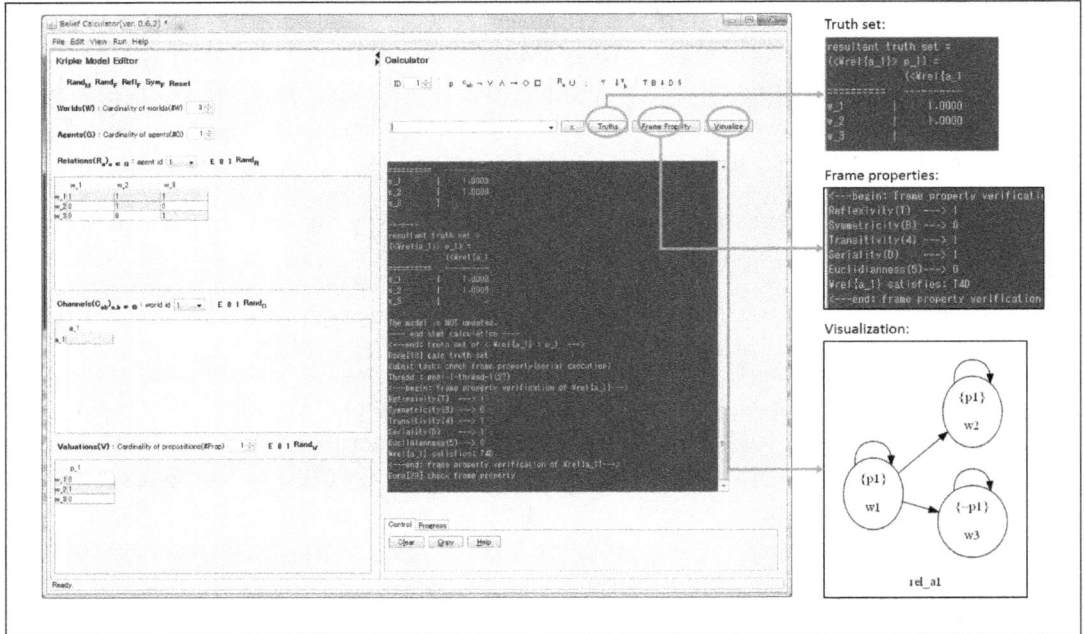

Figure 2: Overview of our implementation

1. Input the cardinality of possible worlds and propositions into corresponding parameter boxes.

2. Input 0 or 1 into each component of matrices for relations, and valuations.[12]

As a matter of practical convenience, each component of matrices works as either a button or a text field. We may either switch the values of matrices 0 and 1 by clicking the component or enter a truth value to the component directly. Each component turns to blue if it has the value 1, white if 0. The colored matrices are helpful since these matrices allow us to recognize some (frame) properties of matrices at a glance (see Section 4.4.1). In addition, there are buttons E, 0, 1, and Rand to set the values of each matrix as unit square matrix, zero matrix, complete matrix and randomly generated matrix, respectively.

Once parameters are entered to the editor side, we can use the calculator side to solve several computation tasks. The calculator has functions which solve tasks of

---

[12]Matrices of channels are used to define communication channels among agents in [6]. In this paper, we leave the matrix to **1**, i.e., the unit square matrix, and this stands for 'every agent has communication channels each other' (cf. Figure 2). Since this is out of focus of the present paper, we can ignore this matrix.

the following kind:

1. Visualization of a Kripke model.

2. Computation of the truth set of a formula.

3. Verification of frame properties listed in Table 1 of a frame

The function for visualizing a Kripke model can be executed by clicking `Visualize` button on the calculator. With the help of 'Graphviz', the function yields and saves a picture of the graphical representation of the model under appropriate directory. Afterward, our software displays the picture on the screen. We will explain details and applications of the other two functions for educational purposes in the following Section 4.3 (computation of a truth set of a formula) and Section 4.4 (verification of frame properties), respectively.

4.3 Computation of Truth Sets and Validity

One of the most basic topics of Kripke semantics for modal logic is to calculate the truth value of a formula at a given world. In connection with this topic, we have the following topics which should be taught to students:

- The truth value of a formula at a given world

- The validity and the invalidity of a formula on a model

In this section, we explain to follow the above topics.

4.3.1 Truth Value of Formula at World

In modal logic, the truth value of a formula is computed at each possible world. In general, we explain to students how the truth value of a formula is computed by the ordinary model-theoretic approach as in Example 2 (Section 2.1). If a given formula is simple and the domain of a given model is small, we can calculate the truth value of a formula visually. This is one of the best points of modal logic. In this approach, we firstly draw a picture of a graphical representation of a Kripke model, and next we calculate the truth value of a formula on the picture.

For example, let us consider a Kripke model $\mathfrak{M}_1$ by $W = \{\, w_1, w_2, w_3 \,\}$, $R = \{\, (w_1, w_1), (w_1, w_2), (w_1, w_3), (w_2, w_2) \,\}$, and $V(p) = W$ (see Figure 3). In the model, $\Box p$ is true at every world. In order to teach the truth of $\Box p$ at a given world on a model visually, we often use the graphical representation of a Kripke model such as a picture of the model $\mathfrak{M}_1$ shown in the left side of the Figure 3. By tracing

131

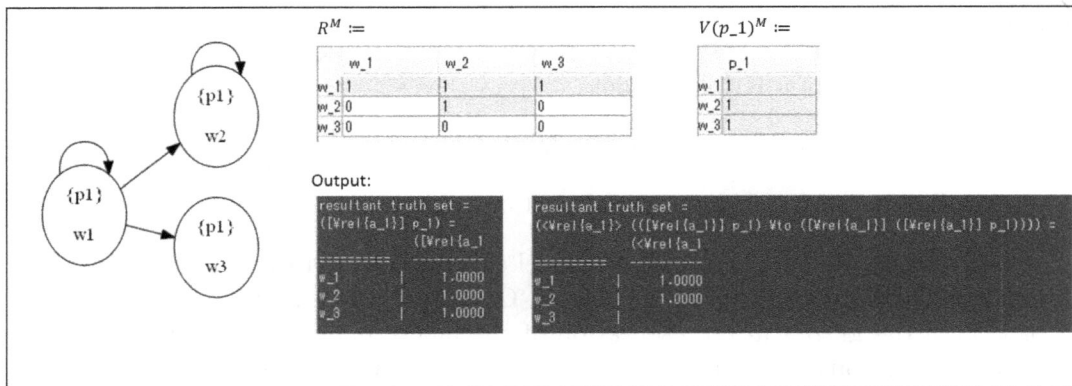

Figure 3: The world w_3 is now 'dead-end'

the links of an accessibility relation from w_1 and w_2, students can obtain the truth value of $\Box p$ at w_1 and w_2, respectively. However, the graphical approach is intuitive but sometimes misleading. For example, some students might be confused how to obtain the truth value of the formula $\Box p$ at w_3. This is because the world w_3 is a 'dead-end,' i.e., a world where we cannot access any world, and so we cannot find any link to the other worlds from the picture. In such a case, we should use the ordinary model-theoretic approach. By the explanation of this approach, students eventually understand why $\Box p$ trivially holds at w_3. Namely, our goal is to show: for all $v \in W$, $w_3 R v$ implies $\mathfrak{M}_1, v \models \varphi$. But, by definition of R, there are no world $v \in W$ such that $w_3 R v$. Therefore, the above implication is vacuously true, and we can conclude that $\Box p$ trivially holds at w_3. However, this proof might be unnatural for some students. In such case, we may also explain the proof by the negation of an assumption. Namely, we assume that $\mathfrak{M}_1, w_3 \not\models \Box p$. Then this is equivalent to:

it is not the case that for all $v \in W$, $w_3 R v$ implies $\mathfrak{M}_1, v \models \varphi$,
iff   for some $v \in W$, it is not the case that $w_3 R v$ implies $\mathfrak{M}_1, v \models \varphi$,
iff   for some $v \in W$, $w_3 R v$ and $\mathfrak{M}_1, v \not\models \varphi$.

Hence, we obtain $w_3 R v$ for some $v \in W$. But there is no world $v \in W$ such that $w_3 R v$ by definition of R, a contradiction. Therefore, the graphical approach sometimes might not work well, and the model-theoretic approach gives us the more rigorous explanation. However, we need to rely on the notion of 'vacuously hold' or the argument by contradiction in this approach.

On the other hand, if we employ our linear algebraic approach to obtain the truth value of the above formula, we do not need to rely on such notions explicitly. As we mentioned in Section 3.1, we can compute the truth value of a formula by an extended truth-table calculation, i.e., a Boolean matrix calculation. In order to teach how to obtain the truth value of $\Box p$ at w_3 by our approach, at first we should show the following matrix representations of R and $V(p)$ to students:

$$R^M := \begin{bmatrix} 1 & 1 & 1 \\ 0 & 1 & 0 \\ 0 & 0 & 0 \end{bmatrix}, V(p)^M := \begin{bmatrix} 1 \\ 1 \\ 1 \end{bmatrix}.$$

Then, we can teach the computation of the truth set of $\Box p$ by:

$$\|\Box p\| = \overline{R^M \overline{V(p)^M}} = \overline{\begin{bmatrix} 1 & 1 & 1 \\ 0 & 1 & 0 \\ 0 & 0 & 0 \end{bmatrix} \overline{\begin{bmatrix} 1 \\ 1 \\ 1 \end{bmatrix}}} = \overline{\begin{bmatrix} 1 & 1 & 1 \\ 0 & 1 & 0 \\ 0 & 0 & 0 \end{bmatrix} \begin{bmatrix} 0 \\ 0 \\ 0 \end{bmatrix}} = \overline{\begin{bmatrix} 0 \\ 0 \\ 0 \end{bmatrix}} = \begin{bmatrix} 1 \\ 1 \\ 1 \end{bmatrix}.$$

Afterward, we can extract the computation of w_3 from the above computation as:

$$\overline{\begin{bmatrix} 0 & 0 & 0 \end{bmatrix} \overline{\begin{bmatrix} 1 \end{bmatrix}}} = \overline{\begin{bmatrix} 0 & 0 & 0 \end{bmatrix} \begin{bmatrix} 0 \end{bmatrix}} = \overline{\begin{bmatrix} 0 \end{bmatrix}} = \begin{bmatrix} 1 \end{bmatrix}.$$

We may also explain that the truth value of $\Box p$ eventually must be true at the dead-end world since $\begin{bmatrix} 0 & 0 & 0 \end{bmatrix} \begin{bmatrix} x \end{bmatrix}$ (where x is a truth value of p at the dead-end world) always returns 1. As we can see above, we can compute the truth set of a formula using our approach easily. On the other hand, in order to obtain the truth set of a formula, in the ordinary model theoretic approach we need to calculate the truth value of the formula for every world.

Furthermore, we may use our software to obtain the truth set of a formula quickly. Our software provides a function to compute the truth set of a formula. Inputs of the function are matrices of a Kripke model and a formula written in TeX style, e.g., 'p ¥land q.' An output is a vector corresponding to the desired truth set. We can compute the truth set of a given formula by clicking the button `Truths` on the calculator. If the computation procedure finishes successfully, a resultant vector appears on a terminal window. See the bottom side of the Figure 3. We can find the vector $\|\Box p\| = 1$ of the truth set. We can also find the more involved computation result in the bottom right side of the figure, i.e., the vector $\|\Diamond(\Box p \rightarrow \Box\Box p)\| = {}^t\begin{bmatrix} 110 \end{bmatrix}$ of the truth set. Since the function yields intermediate computation results, it can be helpful for educational purposes. For example, students can use this function for their self-study. Lecturers can also use this function to provide exercises and write lecture materials. In addition, we can obtain the following solutions from the vector of the truth set of a formula:

1. the truth value of a formula at a world.

2. the validity of a formula on a model.

That is, if the n-th component of the vector is 1, the formula is true at n-th world (item 1). In the right side of the Figure 3, we can find the truth value 1 of a formula $\Box p$ at the row of w_3 of the vector. If the vector is filled with 1, the formula is valid on a model (item 2). Otherwise, the formula is invalid on the model. For the validity and the invalidity, we explain them in Section 4.3.2.

4.3.2 Validity and Invalidity of Formula on Model

In connection with the topic of the truth value of a formula, the validity of a formula on a model is another important topic which should be taught to students. This is because the notion of the validity of a formula is used to explain the notions of a counter-model to a formula, the satisfiability of a formula on a model, and the correspondence between frame properties and (valid) formulas.

In the ordinary model-theoretic approach, we explain that a formula is valid on a model if the formula is true at every world. For example, let us recall the model $\mathfrak{M}_1$ which we used in Section 4.3.1 (see also Figure 3) and verify whether the formula $\Box p$ is valid on the model. After the calculation of the truth value of the formula $\Box p$ for each world, we obtain $\Box p$ is true at every world, and can conclude that $\Box p$ is valid on the model $\mathfrak{M}_1$. During the above calculation, we have to repeat the similar argument.

In the linear algebraic approach, we may explain that a formula is valid if a vector of the truth set of a formula is $\mathbf{1}$, in other words, the vector does not contain 0. For example, through the calculation of the truth set $\|\Box p\|$ on the above model $\mathfrak{M}_1$, we obtain the vector $\mathbf{1}$. Therefore, we can conclude that the formula $\Box p$ is valid on the model. Let us suppose a model $\mathfrak{M}_1'$ by the model $\mathfrak{M}_1$ where $V(q) = \{\, w_2 \,\}$. We can also show that $\Diamond(p \vee q) \leftrightarrow \Diamond p \vee \Diamond q$ is valid on $\mathfrak{M}_1'$. It suffices to show $[\![\Diamond(p \vee q)]\!] = [\![\Diamond p \vee \Diamond q]\!]$ (cf. Section 2.1). Since

$$\|\Diamond(p \vee q)\| = R^M (V(p)^M V(q)^M) = (R^M V(p)^M) + (R^M V(q)^M) = \|\Diamond p \vee \Diamond q\|,$$

we obtain $\|\Diamond(p \vee q)\| = \|\Diamond p \vee \Diamond q\|$, i.e., $[\![\Diamond(p \vee q)]\!] = [\![\Diamond p \vee \Diamond q]\!]$, therefore the commutativity of $\Diamond$ over disjunction is reduced to the distributivity of matrix multiplication. Moreover, we can also show $[\![\Diamond \bot]\!] = [\![\bot]\!]$ by $\|\Diamond \bot\| = R^M \mathbf{0} = \mathbf{0} = \|\bot\|$.

From the notion of the validity of a formula on a model, we should also teach invalidity of that on the model. That is, a formula is invalid on a model if the formula is not valid on the model. In other words, there is some world such that a formula is not true. For example, let us define a model $\mathfrak{M}_2$ by $W = \{\, w_1, w_2, w_3 \,\}$, $R = \{\, (w_1, w_1), (w_1, w_2), (w_1, w_3), (w_2, w_2), (w_3, w_3) \,\}$ and $V(p) = \{\, w_2 \,\}$. The model is the same as the model that we explained in Example 2 (Section 2.1). Then, the formula $\Box p$ is no longer valid on the model $\mathfrak{M}_2$ since $\Box p$ is false at w_1 and w_3. Therefore, $\Box p$ is invalid on the model $\mathfrak{M}_2$. In the model-theoretic approach, we need to find such worlds w_1 or w_3 by the calculation of the truth value of the formula for each world. Although in the linear algebraic approach, we only need to find 0 in the vector of the truth set of the formula, i.e.,

$$\overline{\begin{bmatrix} 1 & 1 & 1 \\ 0 & 1 & 0 \\ 0 & 0 & 1 \end{bmatrix} \overline{\begin{bmatrix} 0 \\ 1 \\ 0 \end{bmatrix}}} = \overline{\begin{bmatrix} 1 & 1 & 1 \\ 0 & 1 & 0 \\ 0 & 0 & 1 \end{bmatrix} \begin{bmatrix} 1 \\ 0 \\ 1 \end{bmatrix}} = \overline{\begin{bmatrix} 1 \\ 0 \\ 1 \end{bmatrix}} = \begin{bmatrix} 0 \\ 1 \\ 0 \end{bmatrix}.$$

Since the vector of the truth set contains 0, we can conclude that the formula $\Box p$ is invalid on the model $\mathfrak{M}_2$. If we use our software, we can check the validity of a formula easily. As we mentioned in Section 4.3.1, we can compute the truth set of a formula on a model, and from the resultant vector we can check the validity of the formula on the model. For example, in Figure 3, the resultant vector of the truth set of the formula $\Box p$ does not contain 0. Therefore, the formula is valid on the model of the figure.

From the invalidity of a formula, we may also explain that a model is a counter-model to the formula. We say that a model $\mathfrak{M}$ is a counter-model to a formula φ if φ is invalid on the model $\mathfrak{M}$. For example, the above model $\mathfrak{M}_2$ is a counter-model to the formula $\Box p$. Of course we can investigate the validity of a formula on the larger model easily. Such investigation can be a good exercise to some students who want to study finite model checking.

4.4 Verification of Frame Properties

As we mentioned in Section 4.3, the truth value of a formula is determined for each possible world. In particular, if the formula contains modal operators, the resultant truth value is affected by the properties of a given accessibility relation. Therefore, it is important to explain topics of various properties of frames to students. In this section, we explain the following topics that should be taught to students:

1. The verification of frame properties of a given frame.

2. The validity of a formula on a frame which satisfies one of the frame properties listed in Table 1.

4.4.1 Frame Properties on Frame

After giving a brief introduction to frame properties of a frame visually, we should explain how to verify them. The ordinary approach to teach the verification of

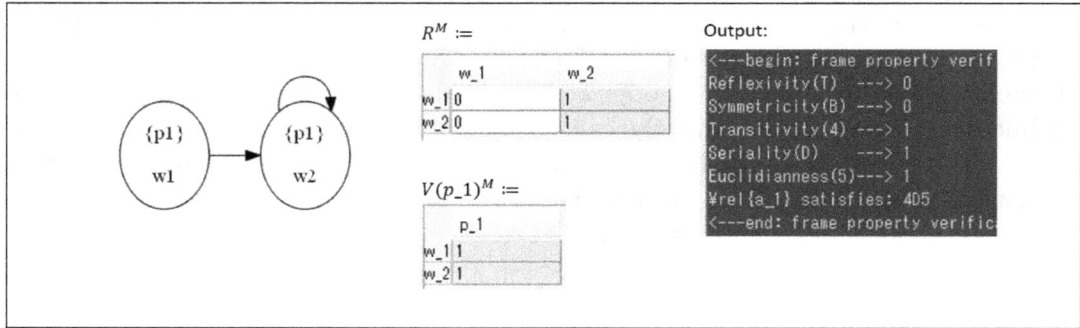

Figure 4: The frame satisfies seriality, transitivity and Euclideanness

frame properties is to use both the visual and the model-theoretic approach. For example, let us define a frame $\mathfrak{F}$ by $W = \{\, w_1, w_2 \,\}, R = \{\, (w_1, w_2), (w_2, w_2) \,\}$ (see Figure 4). The frame $\mathfrak{F}$ satisfies seriality, transitivity and Euclideanness. Since the cardinality of the accessibility relation as a set is enough small, we can use the graphical approach to explain that the frame satisfies the above frame properties. If we show a graphical representation of the frame to students, they might easily realize that the frame satisfies seriality and transitivity. However, it might be difficult to realize if the frame satisfies Euclideanness. In such case, we should switch our explanation to the model-theoretic approach. When we show the frame satisfies Euclideanness, we should check whether the frame satisfies the frame condition of the frame property, i.e., wRv and wRu imply vRu for any $w, v, u \in W$. By definition of the accessibility relation R, we have the following implications:

- $w_1 R w_2$ and $w_1 R w_2$ imply $w_2 R w_2$ ($w = w_1$, $v = w_2$, $u = w_2$).

- $w_2 R w_2$ and $w_2 R w_2$ imply $w_2 R w_2$ ($w = w_2$, $v = w_2$, $u = w_2$).

The other implications, e.g., $w_1 R w_1$ and $w_1 R w_1$ imply $w_1 R w_1$ ($w = w_1$, $v = w_1$, $u = w_1$), trivially hold since the antecedent of the implication is false by definition of R. Therefore, we can conclude that the frame $\mathfrak{F}$ satisfies wRv and wRu imply vRu for any $w, v, u \in W$, i.e., Euclideanness.

If we teach how to verify the frame properties of a frame by our linear algebraic approach, we should show the verification of matrix reformulation of a frame property. Similarly to Example 8 (Section 3.2), we can explain the verification of

Euclideanness of the above frame $\mathfrak{F}$ by $R = {}^t\!RR + R$, i.e.,

$$\begin{bmatrix} 0 & 1 \\ 0 & 1 \end{bmatrix} = \begin{bmatrix} 0 & 0 \\ 1 & 1 \end{bmatrix}\begin{bmatrix} 0 & 1 \\ 0 & 1 \end{bmatrix} + \begin{bmatrix} 0 & 1 \\ 0 & 1 \end{bmatrix}$$

$$= \begin{bmatrix} 0 & 0 \\ 0 & 1 \end{bmatrix} + \begin{bmatrix} 0 & 1 \\ 0 & 1 \end{bmatrix} = \begin{bmatrix} 0 & 1 \\ 0 & 1 \end{bmatrix}.$$

In addition, we may also mention that the more simplified method to verify some frame properties of a frame. We can check whether a given frame satisfies reflexivity, seriality, and symmetricity by the form of a matrix of an accessibility relation of the frame. This is because if the frame satisfies the above three properties, then the matrix of an accessibility relation has the following features:

- Reflexivity: every diagonal component of the matrix consist of 1.

- Seriality: every row contains at least one occurrence of 1.

- Symmetricity: a matrix is a mirror image in the diagonal line.

For example, we can recognize that the matrix R of $\mathfrak{F}$ in Figure 4 satisfies seriality but not reflexivity and symmetricity at a glance.

We may also teach the above matrix computation with our software, which has a function to verify every frame properties listed in Table 1 (Section 3.2). An input parameter of the function is a matrix of an accessibility relation of a frame, and an output is a list of the verification results of each property. After entering the input parameters, it is ready to verify the property. If the dimension of the matrix of an accessibility relation is enough small, we can also use the simplified method to verify some frame properties as we mentioned before. Since the color of each component is blue if it is 1, we can easily recognize whether the matrix satisfies the above features of frame properties. For example, see the matrix of R^M at the center of Figure 4. We can observe that the matrix actually satisfies seriality since every row contains a blue component, i.e., 1. If we wish to use the function of verification of frame properties, we can execute it by clicking the button **Frame Property** on the calculator. If the function finishes successfully, a resultant list will be displayed in a terminal window (see right side of Figure 4). In the list, if the given frame satisfies a frame property, 1 appears at the right side of the name of the property, otherwise, 0 appears. In addition, the name of the modal axiom also appears at the next to the name of the corresponding frame property. At the right side of the Figure 4, we can see that the frame satisfies seriality, transitivity and Euclideanness.

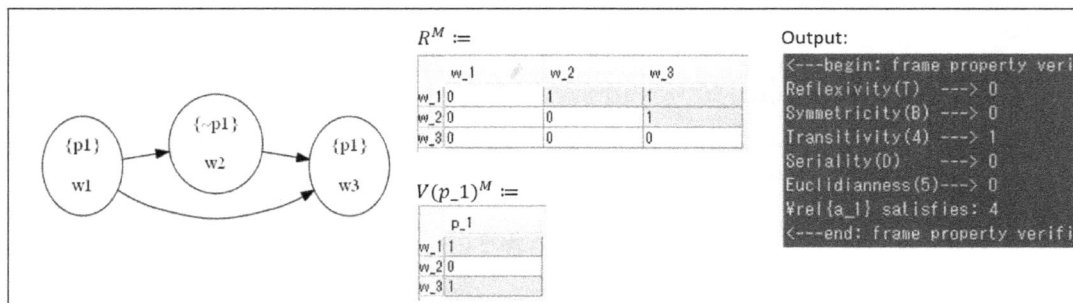

Figure 5: How to satisfy both transitivity and seriality by adding an edge?

Since we can easily manipulate an input matrix of a model by the model editor and obtain the result of the verification of frame properties of the model quickly, for example, we can design or solve the following exercises.

Example 15. Suppose a model of Figure 5 which satisfies transitivity.

1. In order to satisfy both transitivity and seriality, which edge should we add to the model? (answer: add an edge from w_3 to itself.)

2. In order to satisfy Euclideanness, how should we modify the model? (answer: delete every edge from the model.)

3. How to remove every frame property from the model by one edge deletion? (answer: delete an edge from w_1 to w_3.)

If we try to compute possible solutions of the above exercises without supporting tools, we have to compute frame conditions with various changes over and over again. Therefore, our software might be helpful to avoid such efforts.

4.4.2 Frame Properties and Valid Formulas

In this section, we focus our attention on the modal axioms T, B, D, 4 and 5. If we know whether a given frame satisfies some properties, we can determine which formulas are valid. For example, each frame property listed in Table 1 (Section 3.2) has the corresponding formula. If a model satisfies seriality, then the corresponding formula D ($\Box p \to \Diamond p$) is valid on the model. In order to explain the correspondence between frame properties and valid formulas, we use both verification of the validity of a formula on a model in Section 4.3.2 and of frame properties of a frame in Section 4.4.1.

For example, let us recall the frame $\mathfrak{F}$ of previous Section 4.4.1, and define a model $\mathfrak{M}_3$ by $\mathfrak{F}$ and $V(p) = W$ (see Figure 4). The model satisfies seriality, transitivity, and Euclideanness. In addition, the formulas D ($\Box p \to \Diamond p$) for seriality, 4 ($\Box p \to \Box\Box p$) for transitivity and 5 ($\Diamond p \to \Box\Diamond p$) for Euclideanness are valid on the model $\mathfrak{M}_3$, respectively. In order to teach the the correspondence between frame properties and valid formulas smoothly, the above properties are sometimes provided as an assumption. Otherwise, we should start to explain from the verification of all frame properties listed in Table 1 of the frame $\mathfrak{F}$. In this example, we suppose that the above properties are given. Then, we can start our explanation from the verification of the validity of the above formulas. To make our discussion simpler, we focus our attention on the validity of the formula 5 ($\Diamond p \to \Box\Diamond p$) only. Under the model-theoretic approach, we have to check the truth value of the formula 5 for each world. Through the similar discussion in Section 4.3.2, we can eventually conclude that the formula 5 is valid on the model $\mathfrak{M}_3$. Similarly, we may explain the validity of the formula by our linear algebraic approach. By

$$
\begin{aligned}
\|\Diamond p \to \Box\Diamond p\| &= \overline{\overline{\|\Diamond p\|} + \|\Box\Diamond p\|} \\
&= \overline{\overline{R^M V(p)^M} + R^M\left(\overline{R^M V(p)^M}\right)}
\end{aligned}
$$

$$
= \overline{\overline{\begin{bmatrix} 01 \\ 01 \end{bmatrix} \begin{bmatrix} 1 \\ 1 \end{bmatrix}} + \begin{bmatrix} 01 \\ 01 \end{bmatrix} \left(\overline{\begin{bmatrix} 01 \\ 01 \end{bmatrix} \begin{bmatrix} 1 \\ 1 \end{bmatrix}}\right)}
$$

$$
= \overline{\overline{\begin{bmatrix} 1 \\ 1 \end{bmatrix}} + \begin{bmatrix} 01 \\ 01 \end{bmatrix} \overline{\begin{bmatrix} 1 \\ 1 \end{bmatrix}}} = \overline{\begin{bmatrix} 0 \\ 0 \end{bmatrix} + \overline{\begin{bmatrix} 0 \\ 0 \end{bmatrix}}} = \begin{bmatrix} 1 \\ 1 \end{bmatrix},
$$

we can conclude that the formula 5 is valid on the model $\mathfrak{M}_3$. In a similar manner, we can also verify whether the formulas 4 ($\Box p \to \Box\Box p$) and D ($\Box p \to \Diamond p$) are valid on the model, respectively. However, we need to give the effort to calculate matrices, thus we may use our software to check the validity of the formulas quickly.

We should also explain that if a formula which defines a frame property is not valid, then the corresponding frame property is not satisfied in the frame of the model. Remark that we still focus on the axioms T, B, D, 4 and 5. For example, let us define a model $\mathfrak{M}_4$ by $W = \{ w_1, w_2 \}$, $R = \{ (w_1, w_2) \}$ and $V(p) = W$ (see Figure 6). Then the frame of the model $\mathfrak{M}_4$ does not satisfy seriality and Euclideanness since the formulas D ($\Box p \to \Diamond p$) and 5 ($\Diamond p \to \Box\Diamond p$) are no longer valid on the model $\mathfrak{M}_4$, respectively. To show that the frame does not satisfy Euclideanness, we should find a link which violates the frame condition of the frame property. By the ordinary model-theoretic approach, we can find that the following implication does not hold:

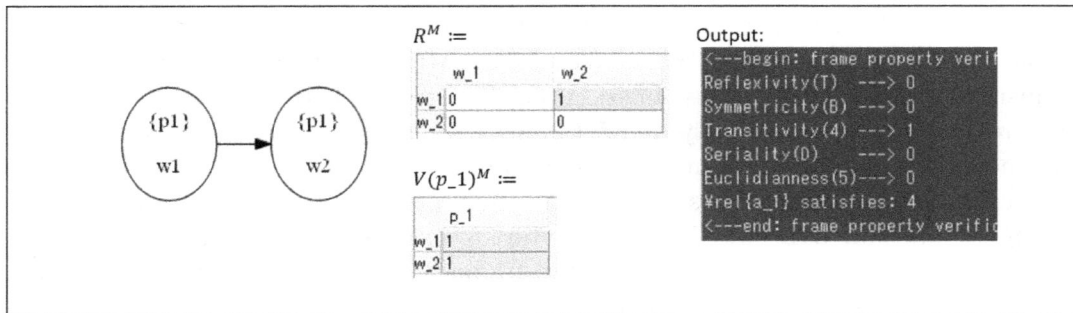

$R^M :=$

	w_1	w_2
w_1	0	1
w_2	0	0

$V(p\_1)^M :=$

	p_1
w_1	1
w_2	1

```
Output:
<---begin: frame property veri
Reflexivity(T) ---> 0
Symmetricity(B) ---> 0
Transitivity(4) ---> 1
Seriality(D)    ---> 0
Euclidianness(5)---> 0
¥rel{a_1} satisfies: 4
<---end: frame property verifi
```

Figure 6: The formula 5 $(\Diamond p \to \Box\Diamond p)$ is invalid and the frame does not satisfy Euclideanness

$w_1 R w_2$ and $w_1 R w_2$ imply $w_2 R w_2$ ($w := w_1$, $v := w_2$, $u := w_2$). Therefore, the frame does not satisfy Euclideanness. The result is also the same in the linear algebraic approach by $R \neq {}^t RR + R$, i.e.,

$$\begin{bmatrix} 0 & 1 \\ 0 & 0 \end{bmatrix} \neq \begin{bmatrix} 0 & 0 \\ 1 & 0 \end{bmatrix} \begin{bmatrix} 0 & 1 \\ 0 & 0 \end{bmatrix} + \begin{bmatrix} 0 & 1 \\ 0 & 0 \end{bmatrix}$$

$$= \begin{bmatrix} 0 & 0 \\ 0 & 1 \end{bmatrix} + \begin{bmatrix} 0 & 1 \\ 0 & 0 \end{bmatrix} = \begin{bmatrix} 0 & 1 \\ 0 & 1 \end{bmatrix}.$$

The above calculation seems easy but it takes a bit of our time. Hence, we may use our software to verify which frame properties are satisfied on the frame quickly.

At the end, for example, we can design the following exercise with the help of our software.

Example 16. Let us define a Kripke model by $W = \{ w_1, w_2, w_3 \}$, $R = \{(w_1, w_1), (w_1, w_3), (w_2, w_3), (w_3, w_1), (w_3, w_3)\}$ and $V(p_1) = \{ w_1 \}$ (see Figure 7).

1. Enumerate satisfying frame properties (listed in Table 1). (answer: the frame satisfies seriality and Euclideanness.)

2. Verify whether the formula 5 $(\Diamond p_1 \to \Box\Diamond p_1)$ is valid on the model. (answer: 5 is valid on the model.)

3. Let us delete an edge from w_3 to w_1. Afterward, verify again the formula 5 on the model. (answer: 5 is invalid on the model.)

4. Verify whether the frame satisfies Euclideanness. (answer: the frame does not satisfy Euclideanness.)

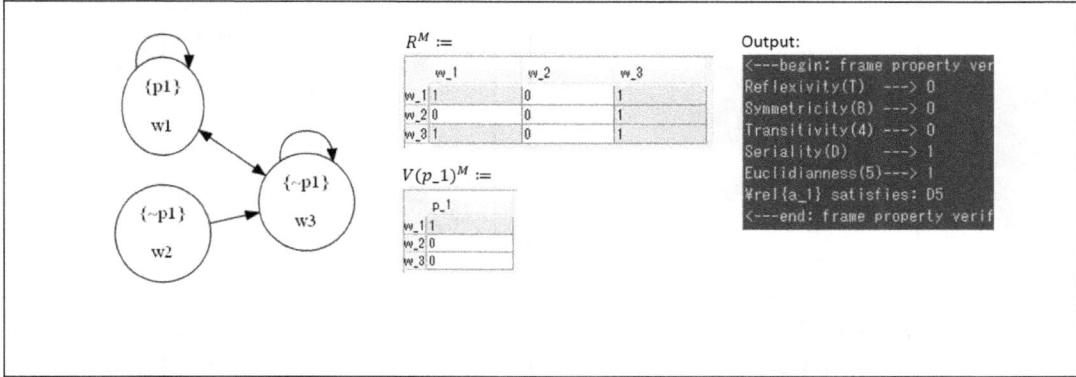

Figure 7: Which frame properties are satisfied on the frame?

When we teach modal logic, we sometimes need to consider involved exercises as above. With the help of some supporting tools, we can avoid our efforts by hand and eliminate human errors from our teaching materials.

4.5 Feedbacks from Students

In order to obtain feedbacks from students, we have held a small seminar to teach elementary topics of modal logic using our approach. The participants of this seminar were 15 graduate students from our university. For reference, we have opened our lecture material and feedbacks from students for the public (see also Figure 8).[13] In the lecture, we taught the following topics to the students:

1. Truth-table calculation of propositional logic, how to read modal operators, how to define formulas, and the dual-definition of modal operators.

2. A graphical representation of Kripke model, linear algebraic reformulation of Kripke semantics, and computation of truth sets and validity of a given formula on a model.

3. Matrix representation of the five frame properties in Table 1 (reflexivity, symmetricity, transitivity, seriality and Euclideanness), verification of them, and the correspondence between these frame properties and formulas (T, B, D, 4 and 5).

These topics are selected from teaching topics explained in Sections 4.1, 4.3 and 4.4. We used Examples 2, 3 and 8, and examples of the truth value of a formula $\Box p$ at

---

[13]http://cirrus.jaist.ac.jp:8080/soft/ttl

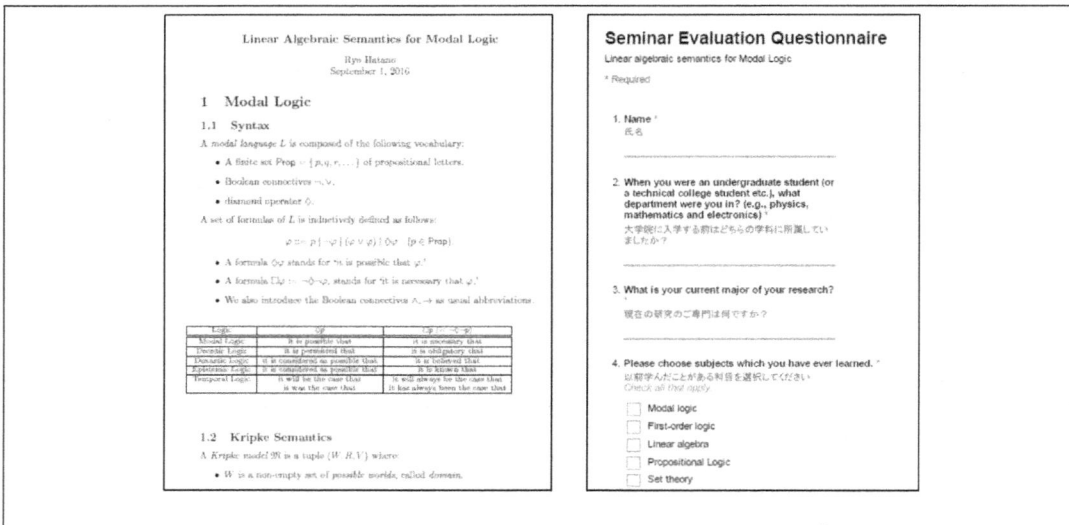

Figure 8: Our lecture material and questionnaire form

a dead-end world in Section 4.3.1 and the verification of Euclideanness property in Section 4.4.1. In order to compare our linear algebraic approach with the ordinary model-theoretic approach, we also provide a short explanation of the model-theoretic approach for each topic. Finally, we demonstrated our supporting software.

After the seminar, we have conducted a survey using a questionnaire form (see the right side of Figure 8). The questionnaire form consists of the following items:

1. Past and current affiliation, past and current major of research and subjects which he/her has ever learned.

2. Levels of understanding of the topics and his/her intriguing topics.

3. Preferred approach to learn modal logic.

4. Effectiveness of our approach and supporting software to learn modal logic.

We have collected 15 completed questionnaire forms, and the results and opinions can be summarized as follows. For item 1, we found that 14 students have already learned linear algebra, and 6 students have never learned first-order logic (see the left diagram of Figure 9).[14] In what follows, we regard the latter 6 students (i.e.,

---

[14]In the questionnaire, we also asked a question that whether students know modal logic and set theory. As a result, we found that 6 students and 4 students have already learned modal logic and set theory, respectively. Since these students also have already learned first-order logic, we

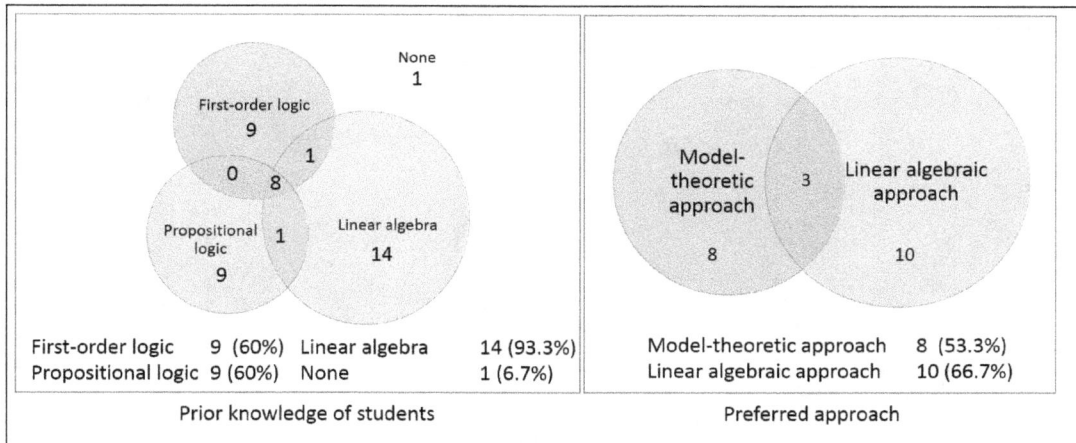

Figure 9: Prior knowledge of students and their preferred approach

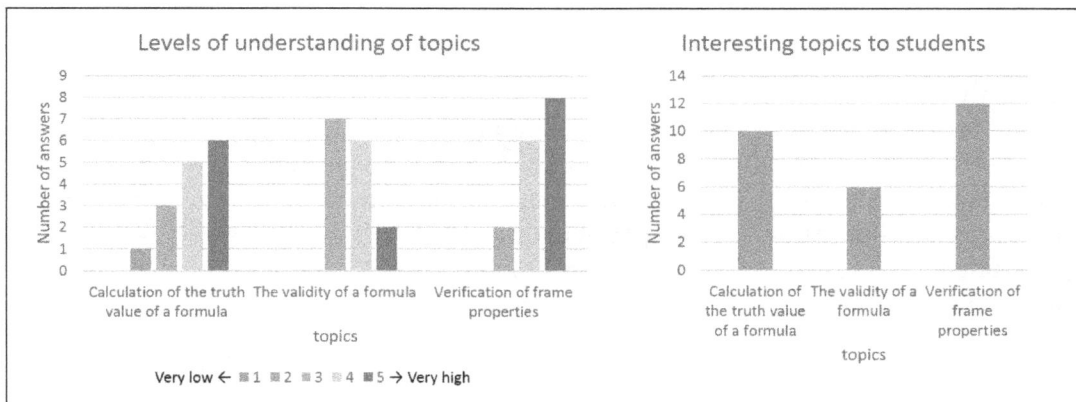

Figure 10: Interesting topics to students

one-third of the participants) as our target students since we taught truth-table calculation at the beginning of the seminar. For item 2, most of our students answered that they could understand the topics of the seminar (see the left graph of Figure 10). In particular, the topic of the verification of frame properties using matrices attracts the interest of 12 students (see the right graph of Figure 10). For item 3, the results were different from our expectation. Our approach was bit preferred than the model-theoretic approach; 7 students preferred to use linear algebraic approach, 5 students preferred to use the model-theoretic approach and 3 students preferred to

---

have merged them into the same group of students who know first-order logic in the left diagram of Figure 9.

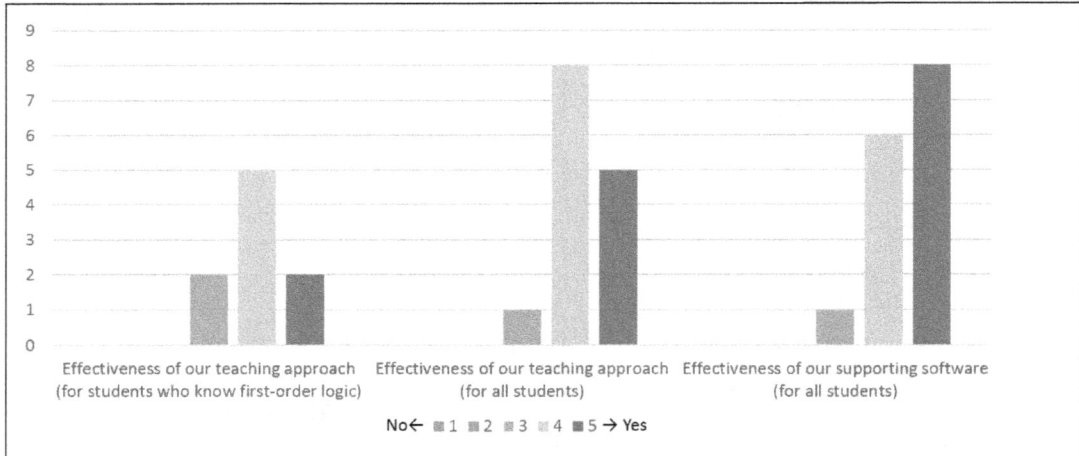

Figure 11: Effectiveness of our teaching approach and software to learn modal logic

use both approaches (see the right diagram of Figure 9). In particular, 5 students of our target preferred the linear algebraic approach. Their opinion is that simple matrix calculations allow them to understand the elementary notion of modal logic since they are not familiar with set theory and felt difficult to understand the model-theoretic treatment of Kripke semantics. This result indicates that our target students could learn elementary part of modal logic using our approach. On the other hand, 7 students who have already learned first-order logic were divided into two groups; 4 students preferred model-theoretic approach and 3 students preferred both approaches. But these 7 students also answered that they could deepen their understanding of modal logic from the linear algebraic perspective (see the left graph of Figure 11). For item 4, most of our students are agreed that our approach is effective to learn modal logic (see the center graph of Figure 11). They also answered that our software can be helpful to their study since they wished to avoid involved calculation of matrices (see the right graph of Figure 11).

As a result, students eventually got a positive impression to learn modal logic using our approach. The above feedbacks indicate that our approach must be efficient for both our target students and those who have already learned first-order logic. The result also indicates that our approach has the potential of expanding the range of our target students. For example, in Japan, many high-school students learn basic calculations of real-valued matrices. Therefore, we may teach elementary topics of modal logic and graph theory to advanced high school students in terms of matrices.

5 Conclusion

We have explained a linear algebraic approach to teach modal logic to students. Based on Fitting's linear algebraic reformulation of Kripke semantics of modal logic, we can represent an accessibility relation R by a Boolean matrix, and a valuation $V(p)$ of an atomic variable p by a Boolean column vector, provided the cardinality of the domain is finite. Then we can calculate the truth set of a formula by calculations over these matrices.

Moreover, we can also verify the frame properties in terms of matrices. In order to obtain a linear algebraic reformulation of frame properties, we have shown two types of correspondence between matrix reformulation of frame properties and corresponding modal axioms. Based on the idea of Lemmon-Scott axioms and Geach axioms, we can capture all of the five well-known frame properties, i.e., reflexivity, symmetricity, transitivity, seriality and Euclideanness. As a result, we can regard the above matrix calculations of Kripke semantics as an extended version of truth table calculation. Furthermore, this extension allows us to capture some restricted form of quantifications (in Kripke semantics) without bound variables of first-order logic.

Our target students are those who have prior knowledge of linear algebra and propositional logic. They can learn modal logics based on their acquired knowledge without learning first-order logic. In addition, students who have already learned first-order logic, can also deepen their understanding of the subject from a different perspective. In order to claim this, we have explained which topics of modal logics can be taught to students using our approach and why the approach is helpful for educational purposes. In order to avoid involved calculations on matrices, we have introduced our supporting software. Finally we have taught some elementary topics of modal logic to our students using our approach, and collected feedbacks from them. The feedbacks indicates that our approach must be efficient for both our target students and the ordinary target students of a course of modal logic.

A further direction of this study will be to use a similar approach to teach advanced topics of modal logics. In particular, we may expand our mono-modal syntax into multi-modal one to cover, e.g., description logic [1] and dynamic epistemic logic [17]. As for description logic, a family of *roles* (say "has a child") generates both box-type and diamond-type modal operators. Therefore, we can capture the semantics of these operators by a set of the corresponding adjacency Boolean matrices to the roles. We can also cover some topics of dynamic epistemic logics [17] by our approach, where multi-modal operators are employed for describing agents' knowledge or beliefs. For example, [6] has presented a linear algebraic semantics of dynamic epistemic logic for a multi-agent communication system, with the help of van Benthem and Liu's idea of relation changer [16]. A key idea of the notion of relation changer is that a dynamics of beliefs may be described as a *program term* in propositional dynamic logic, i.e., a program term constructed from atomic programs by composition, non-deterministic choice, and test. Since we can represent these three program constructors as matrix operations, we may also apply our approach to such dynamic epistemic logic. The other topics, such as soundness and completeness theorem of modal logics, still remain for further studies.

References

[1] Franz Baader, Deborah L McGuinness, and Daniele Nardi. *The Description Logic Handbook: Theory, implementation, and applications*. Cambridge University Press, 2003.

[2] Rudolf Berghammer and Renate A. Schmidt. Investigating finite models of non-classical logics with relation algebra and relview. In *Proceedings of the 2006 International Conference on Theory and Applications of Relational Structures As Knowledge Instruments - Volume 2*, TARSKI'02-05, pages 31–49, Berlin, Heidelberg, 2006. Springer-Verlag.

[3] Max J Cresswell and George Edward Hughes. *A new introduction to modal logic*. Routledge, 2012.

[4] Melvin Fitting. Bisimulations and boolean vectors. In *Advances in Modal Logic*, pages 97–126. King's College Publications, 2003.

[5] Akira Fusaoka, Katsunori Nakamura, and Mitsunari Sato. On a linear framework for belief dynamics in multi-agent environments. In *Computational Logic in Multi-Agent Systems*, pages 41–59. Springer, 2007.

[6] Ryo Hatano, Katsuhiko Sano, and Satoshi Tojo. Linear algebraic semantics for multi-agent communication. In *Proceedings of the 7th International Conference on Agents and Artificial Intelligence*, volume 1, pages 172–181, 2015.

[7] K.H. Kim. *Boolean matrix theory and applications*. Monographs and textbooks in pure and applied mathematics. Dekker, 1982.

[8] E.J. Lemmon and D.S. Scott. *An Introduction to Modal Logic: The Lemmon Notes*. American philosophical quarterly monograph series. B. Blackwell, 1977.

[9] Churn-Jung Liau. Matrix representation of belief states: An algebraic semantics for belief logics. *International Journal of Uncertainty, Fuzziness and Knowledge-based Systems*, 12(05):613–633, 2004.

[10] Ewa Orlowska. Relational formalisation of nonclassical logics. In Chris Brink, Wolfram Kahl, and Gunther Schmidt, editors, *Relational Methods in Computer Science*, pages 90–105. Springer Vienna, Vienna, 1997.

[11] J. A. Plaza. Logics of public communications. In M. L. Emrich, M. S. Pfeifer, M Hadzikadic, and Z. W. Ras, editors, *Proceedings of the 4th International Symposium on Methodologies for Intelligent Systems*, pages 201–216, 1989.

[12] Gunther Schmidt and Thomas Ströhlein. Relations and graphs: Discrete mathematics for computer science. *EATCS Monographs on Theoretical Computer Science*. Springer-

Verlag, 1993.

[13] Alfred Tarski. On the calculus of relations. *The Journal of Symbolic Logic*, 6(3):73–89, 1941.

[14] Satoshi Tojo. Collective belief revision in linear algebra. In *Computer Science and Information Systems (FedCSIS), 2013 Federated Conference on Computer Science and Information Systems*, pages 175–178. IEEE, 2013.

[15] Johan van Benthem and Guram Bezhanishvili. Modal logics of space. In *Handbook of spatial logics*, pages 217–298. Springer, 2007.

[16] Johan van Benthem and Fenrong Liu. Dynamic logic of preference upgrade. *Journal of Applied Non-Classical Logics*, 17(2):157–182, 2007.

[17] Hans van Ditmarsch, Wiebe van der Hoek, and Barteld Pieter Kooi. *Dynamic epistemic logic*, volume 337. Springer, 2007.

Received 11 October 2016

The New Trivium

Patrick Blackburn

Section for Philosophy and Science Studies, University of Roskilde, Denmark
patrick.rowan.blackburn@gmail.com

Abstract

The medieval trivium consisted of logic, grammar and rhetoric. It was the 'linguistic' component of medieval education as opposed to the more mathematical quadrivium. Over the last thirty years or so something worth calling the New Trivium has emerged. This paper discusses what it is, and why it is important, but is for the most part concerned with a pedagogical issue: how best to teach the New Trivium to humanities students with little or no background in logic.

Keywords: Trivium, Quadrivium, Logic, Critical thinking, Logic education, Problematisation, Humanities

1 Introduction

Roughly speaking, by the New Trivium I mean certain applications of logic to fields such as linguistics, social and cognitive psychology, knowledge representation, artificial intelligence, and theoretical computer science. This is not very precise, but later in the paper I'll explain why I believe that there is something here worth isolating and calling the New Trivium. Nonetheless, for the most part this paper is not so much concerned with what the New Trivium is, or even why I believe it to be important. Instead, it's mostly about how to teach it, and in particular, how to teach it to students with very little (if any) logical background.

Indeed, this paper is really about *teaching logic without teaching logic*. This may sound paradoxical, or whimsical, but I don't intend it to be either. Rather, the expression "teaching logic without teaching logic" is my attempt to point to a pedagogical gap. At present, New Trivium subjects are mostly taught at graduate

---

I am grateful to my RUC colleagues Camelia Elias, Kasper Eskildsen, Klaus Frovin Jørgensen, Prem Poddar and Irina Polyanskaya who have taught me many new things about teaching.

schools, often specialised summer schools, to students with some technical background. I would like to make this body of work accessible to undergraduate humanities students, including those with little (if any) knowledge of logic. Quixotic this may be; paradoxical it is not.

There is an autobiographical component to the paper. While I have only recently started talking of the New Trivium, the phrase is my attempt to name whatever it was that I learnt as a PhD student from the various research communities around me,[1] and which I have since spent much of my professional life trying to articulate. I have taught New Trivium material at many different levels (undergraduate, graduate, postgraduate) to many different specialist audiences (linguists, computational linguists, computer scientists, and philosophers) and to highly interdisciplinary audiences too. But for the past five years I have been teaching at the humanities faculty at Roskilde University (RUC). This has been one of the most stimulating pedagogical experiences of my life, but also, for reasons I shall now explain, one of the most challenging when it comes to logic education, especially for the topics that I am lumping together under the New Trivium banner.

RUC is Danish university which emphasises student-led project work and interdisciplinarity. At the start of their first semester, humanities students are required to form project groups (typically with 4-6 members) and work on a self-chosen research topic. They do the same at the start of their second and third semesters too. All three projects are worth 15 ECTS credits, and each is assigned an academic supervisor. Much of my time is spent supervising such groups.

Interdisciplinarity is hardwired into RUC's Bachelors and Masters programs. During their first three semesters (that is, while writing the projects just mentioned) students are not attached to any particular department but to the entire humanities faculty.[2] During this time they do not follow specialised courses in (say) history or Danish, but instead follow four wide-ranging lecture courses covering a range of humanities (and indeed, social science) topics. They also participate in more intensive courses and workshops on research methodologies and the role of theory in the humanities. Only in their third semester do they begin to attend the regular classes offered in individual disciplines (for example philosophy, history, or Danish). And the final touch is this: in order to gain the Bachelors degree, every student is required to study *two* such disciplines, and typically both these disciplines will

---

[1]I was a PhD student at the Centre for Cognitive Science at Edinburgh, where I was surrounded by computational linguists, formal semanticists, cognitive psychologists, and logicians.

[2]This is done by assigning them all to different 'houses'. The reader who thinks of these in terms of Gryffindor, Hufflepuff, Ravenclaw and Slytherin has got it pretty much right. House-based student project work has something of the flavour of the Oxbridge college-based tutorial system.

also be studied at Masters level.[3] That is (to use the American terminology) RUC students are required do a double major. It's also worth remarking that the student-led project work mentioned above does *not* stop after the third semester; it continues unbroken until the Masters thesis. Typically 50% of the time spent on both major subjects (at both Bachelors and Masters level) is project based, and just as in the first three semesters, students have considerable freedom in their choice of topics.

This is an unusual system, at least for a European university. Indeed, when I describe it to fellow academics it tends to give rise to either extremely enthusiastic or highly skeptical reactions. I am firmly in the enthusiastic camp. I find RUC's teaching system innovative and exciting, and a major motivation for my move to Denmark was to be a part of it. But I did not outline RUC's teaching system in order to discuss its advantages and disadvantages (that would require a separate paper). Rather it was to make clear that the questions I posed at the start — *how to teach logic without teaching logic, and how to bring humanities students face-to-face with the New Trivium* — are a concrete issues for me. I teach in the Philosophy and Science Studies Section of the humanities faculty. How can I best teach logic in general, and the types of logic relevant to the New Trivium in particular, given the academic setup just described?

Let me make some of the difficulties explicit. The first is fairly obvious: there's simply not enough time. Roughly 50% of students' time is devoted to student-led project work, and courses in the individual disciplines only begin in the third semester. Moreover, even when these courses begin, the remaining time available for lecture courses is split between two disciplines, as each student does a double major. And roughly 50% of the time in each discipline is reserved for project-based work anyway, and here the students' choices reign supreme. The upshot is that each discipline has to strip its compulsory courses down to the essentials, and omit anything not belonging to this core. Again, I applaud: it's a great way to strip away intellectual fat. Unfortunately, it also means that we do not have an introductory logic course in the Philosophy and Science Studies Section.[4]

The second difficulty, however, concerns issues of motivation, and these run deeper. As is probably clear, even from the brief remarks made above, appreciating what the New Trivium offers presupposes a certain sophistication concerning what modern logic is and what it can bring to the study of such subjects as language, cognition, knowledge and interaction; such sophistication is not something

---

[3] As in most European countries, the majority of university students in Denmark complete a five year program consisting of a three year Bachelors degree followed by a two year Masters degree.

[4] The natural sciences faculty, however, offers a Masters course called *Fundamentale matematiske strukturer* which covers some set theory, and a Bachelors course called *Logik og diskret matematik*, which introduces propositional and predicate logic and tableaux-based proof methods.

that undergraduate students (in any subject) can reasonably be expected to have. Moreover, what the New Trivium offers can often seem alien and uninteresting, especially to humanities students. For a start, many of its insights are hidden behind a mask of mathematical formalism, which humanities students tend to dislike, or find intimidating. But this problem, real though it is, may be the most minor one. More pressing is the simple fact that RUC students have a great deal of freedom in choosing what they write their projects on. Students who choose philosophy often do so because they have been inspired by major philosophical figures, ranging from Baruch Spinoza to Judith Butler (Danish students have often gained some background in philosophy from their school studies) and they quite understandably want to devote their precious project time to studying their works in depth. Or they may have chosen philosophy for what it can tell them about contemporary social and political issues; projects on such topics as refugees, issues in feminism and gender, or the relevance of neuroscience to ethics, are fairly typical here. Or they may have chosen philosophy simply because it fits in well with their other major subject which (at least until recently) could be pretty much anything offered in any department in the university. Given all this, logic is easy to overlook and usually is.

What follows are reflections prompted by this situation. I'll discuss a couple of logic-related courses I have given, noting what went right and what went wrong. I'll give further details about the New Trivium, explain why I think it is important, and try to place it in historical context. Above all, I'll try to explain what I think the next step should be, for being forced to think seriously about how to teach logic without teaching logic has taught me a lot. And although it was the special environment of RUC that led me there, what I learnt may be of wider interest.

2   Critical Thinking

I'll start by discussing a course in critical thinking that I first taught in 2013. I do so for two reasons. The first is simply this: many universities do not teach specialised logic courses, but courses in critical thinking are common. In a sense, such courses offer a simple answer to the question of how to teach logic without teaching logic, though it is an answer driven almost totally by practical concerns. My second reason, however, will lead to a central theme of the paper. By the time I taught the critical thinking course I had already taught a course explicitly devoted to New Trivium topics. I'll discuss this (in key respects, unsuccessful) course in some detail later. For now I'll simply say that the experience made me think harder about how to motivate logic-related matters for humanities students, and I approached the critical thinking course with the issue of motivation firmly in mind.

A contemporary course in critical thinking at an Anglo-Saxon university is likely to cover some combination of the following topics: the distinction between inductive and deductive logic, with deductive logic typically being represented by examples of syllogistic reasoning and some propositional logic. Little if any formalism will be used: the aim is to teach the student that there are different types of reasoning, and that there are well understood patterns of correct and incorrect inference. Students are expected to learn to distinguish good from bad reasoning in various kinds of texts, such as newspaper articles and academic papers. Many critical thinking textbooks contain extensive lists (with examples) of logical fallacies, and many (when discussing inductive logic) may well dig deeper into statistics and probability theory. For example, many discuss the perils of ignoring base probabilities (how likely is it that you *really* have that horrible disease you just tested positive for?) and introduce basic Bayesian inference.

There are many textbooks on critical thinking available (there's a big market) and lots of free online material. Some, is dull and formulaic. Others approach the basics with wit, charm and clear writing. Yet others approach critical thinking in a deeper and more systematic way, and take care to address contemporary issues, such as how to get reliable information off the internet (if you have ever supervised projects with students just out of school, you'll appreciate how important this is). Finally, some books dig deeper into the foundations of critical thinking. A central question here is how to get to grips with the logical structure of real texts in a way that makes it possible to assess their argumentation. This is difficult to do, and there is a long research tradition devoted to it; for example, Stephen Toulmin's *The Uses of Argument* [28] is an influential early monograph on this, and a more recent textbook, Alec Fisher's *The Logic of Real Arguments* [10] introduces a novel method of teaching such skills and applies it to some challenging examples.

Nonetheless, recent research suggests that there may be a problem at the heart of critical thinking. Quite simply, there seems to be evidence that such courses are not particularly effective; see, for example, Willingham [32]. They attempt to teach informal logical skills, but it is unclear whether such courses have a lasting effect.

Why so? Because human reasoning skills seem grounded in particular contexts. A historian learns logical and critical thinking in the (long) process of becoming a historian, a chemist learns as part of the process of becoming a chemist. But it is unclear how well thinking skills learnt in this way transfer to other domains. It may be that some disciplines give rise to transferable skills better than others (I have heard mathematicians, computer scientists, and philosophers all argue that their fields are excellent — indeed the very best! — setting for learning critical thinking). Be that as it may, we are all aware of how often experts in one field come badly unstuck in another. And yet the central claim of traditional critical thinking courses is precisely that they teach, in a clear, digestible (and above all) *transferable* form, an important set of domain independent reasoning skills. It is unclear whether such claims are substantiated.

It occurred to me that this difficulty might be an excellent way to *motivate* the course. The point is this. Many courses in critical thinking are not really courses about critical thinking as such — rather they are essentially courses presenting a number of critical thinking techniques, together with lots of practical examples. But few courses and textbooks address in much depth what critical thinking is, why it is desirable, and why it is that (even with the very best intentions on our part) we can go so spectacularly wrong. And yet there is a fascinating literature which addresses all these issues (and many more of relevance) in depth.

The assigned course reading included parts of Dewey's *How We Think* [7], a classic text by one of the 20th century's most influential philosophers and pedagogical theorists, a text very much on the *why* rather than the *how* of critical thinking. I also assigned parts of Kahneman's *Thinking, Fast and Slow* [16] which presents and extends his joint work with Amos Tversky in an engaging manner. Another valuable book was Cordelia Fine's *A Mind of its Own* [9], together with some research papers on cognitive biases. Reading the Dewey text gives students a deeper insight into what critical thinking might be, and what role it has to play in education. The texts by Kahneman, Fine and others made students aware of the wild cognitive whirlwinds, the idiot winds, that blow inside our heads.

I opened the course by asking the students to attempt a number of well known puzzles, including the Wason Four Card Task, and Kahneman and Tversky's Linda the Bank Teller Scenario. The results were every bit as bad as expected. Immediately afterwards, we discussed Willingham [32], the paper mentioned above which suggests that teaching critical thinking may well be a waste of time. In short: the first session was devoted to making the students acutely aware that informal logical thinking is difficult, and that maybe there wasn't really all that much that we could do about it in the short space of the course!

The students liked this. It challenged them. In subsequent sessions we alternated between working through material from standard critical thinking textbooks, and discussing the Dewey, Kahneman, Fine (and other) material. And throughout the course I tried to get the students to ask what critical thinking was, and to understand that it was uninteresting (and probably not particularly useful) to think of it as learning how to bandage logical wounds. Bandages fall off. The deeper point is to understand where the wounds come from, and to be on the lookout for new ones.

To put it another way: *I taught critical thinking by problematising the notion of critical thinking.* That is: I tried to teach my students to think critically by getting them to think critically about critical thinking. I don't know whether this approach has more long-term effect that a more standard approach, but it at least had the merit of treating my students as students of a genuine subject (namely: critical thinking) rather than as consumers of some critical thinking product.

But whether the students learnt something valuable or not, I certainly did. Teaching the central concept of the course (in this case, critical thinking) by problematising it clearly struck a chord. I started to think more systematically about how I could introduce logic in a similar way. And not just any logic; I wanted to teach the New Trivium.

3 The New Trivium

The medieval trivium consisted of logic, grammar and rhetoric. It was the 'linguistic' part of medieval education and paved the way for the more mathematical quadrivium of arithmetic, geometry, music and astronomy. But what is the New Trivium? Consider the courses typically taught at a European Summer School for Logic, Language and Linguistics (ESSLLI).[5] All three components of the original trivium are clearly visible — but in brave new form.

For a start, there is grammar. Natural language semantics is usually well represented, often under the name formal semantics to emphasize the heavy use made of logic. Sometimes the logical ideas involved are long-established; Church-Henkin style type theories, for example, are a staple tool. Often, however, they are new: Discourse Representation Theory (see [17]) and other 'dynamic' approaches to semantics are important examples. And then there's syntax. Richard Montague famously insisted that syntax was interesting only as a prelude to semantics — but the necessary syntactic and phonological 'preludes' (if indeed they are mere preludes) are all taught at ESSLLIs too. All in all, the grammar component of trivium is clearly present, and indeed represented in a variety of forms: theoretical, computational, symbolic and statistical.

What about rhetoric? Here, perhaps, the link is looser, but it is present nonetheless. Rhetoric is the study of certain practical uses of language, in particular the arts of persuasion and effective communication. But the word 'art' here is noteworthy. One of the most interesting components of what I call the New Trivium is contemporary work on *pragmatics*, which gives a much broader perspective on language use. The conceptual link with rhetoric is, I think, clear — but contemporary pragmatics (and related areas such as the study of discourse and dialogue) represent a fundamental step forward in our understanding of language. Modern

---

[5]Or at a North American Summer School for Logic, Language and Linguistics (NASSLLI).

pragmatics has a far wider range of concerns than classical rhetoric.[6] Ideas from social reasoning, AI-style planning, and game-theory give contemporary pragmatics a solid theoretical basis, and its border with modern semantics is an active area of investigation. The science underlying the art is becoming visible.

Gluing these topics together is logic. Or rather logics. For ESSLLIs over the years have offered courses in many kinds of logic: modal logics, epistemic logics, description logic, dynamic logics, default logics, non-monotonic logics, partial and multi-valued logics, fuzzy logics — the list goes on. But what is most worth noting is not the items on the list, but where they come from. Many are logics that arose from practical applications, often in AI and computer science (default and description logics are obvious examples). Others are logics that bridge several areas of investigation: epistemic logic, which has roots in philosophy and computer science is a good example. To put it another way: logic is not a pre-existing glue applied from the outside. It is a glue that is being actively developed and redeveloped as part of ongoing investigations into novel research areas. The remarkable development of Dynamic Epistemic Logic (see [29]) over the last decade or so is a good illustration of this process at work.

In short: logic is changing these areas, and these areas are changing logic. But how did this happen? A little historical context may clarify. From antiquity until the 19th Century (let's say, until John Stuart Mill) logic was seen as relevant to what we now call linguistic and cognitive themes. For logic has always had high ambitions; many of the themes investigated by medieval logicians still have contemporary resonance. Nonetheless, for most of its history, logic's reach exceeded its grasp. It was underpowered for the tasks it set itself.

All this changed with the work of Frege and Cantor. Frege gave us modern logical syntax with its variables and variable binding; Cantor gave us the set theory that would eventually provide the setting for model-theoretic semantics. And logic entered what we might call its first mathematical phase. One of the key themes motivating this phase was that logic might provide a foundation for mathematics. The work of Gödel in the early 1930s largely killed this dream, but it also led the way towards one of logic's most enduring legacies: the creation of a robust notion of computation in the work of Church and Turing.

The insights of the 1930s also led to what we might call logic's second mathematical phase. Here the pioneering figure is Tarski: the syntax-semantics distinction is clearly drawn, the concept of a model clearly emerges, and the relationship of logic

---

[6]While this claim is probably true, it is worth emphasizing that classical rhetoric had a far wider scope and a more central role to play in classical thought than the (often somewhat dismissive) contemporary use of the term 'rhetoric' might suggest; see Chapter 1 of [26].

and mathematics is inverted: far from being a foundation for mathematics, logic becomes a branch of mathematics. The 1960s, with its independence proofs and the invention of non-standard analysis were a golden era for second phase work.

Summing up: much of logic from the time of Frege and Cantor until the 1960s was mathematical in nature and in its areas of application. More irreverently we might say: it is the story of the 90lb weakling who entered the gym and emerged as a Charles Atlas figure, muscles bulging. And this is significant because, sometime around the end of the 1960s, logic started to rediscover its pre-Fregean concern with language and cognition. But now it packed bigger punch.

And it was around then that the interdisciplinary work that was to form the New Trivium began to emerge. First, and perhaps foremost, logic and language become more deeply intertwined. At least two lines of work are significant here. One is Richard Montague's development of formal semantics (see [22]), which showed the power of logical methods in the study of language, and opened the door to computational approaches to semantics. Another is the link between pragmatics and logic. Here, perhaps, the key figure is Paul Grice (see [13]). His work was largely informal, but the connections between his ideas and various kinds of epistemic and dynamic logics have become clearer in subsequent years.

Which leads us to a second thread: the emergence of epistemic logic. Here the pioneering work is due to Jaako Hintikka ([14]); his work would later feed into theoretical computer science, and blend with dynamic approaches to semantics and work on planning from the AI community. And this swiftly leads us to a third thread: the early link forged between logic and AI. A key paper here is McCarthy and Hayes [19], which drew topics from philosophical logic and transformed them into tools for thinking about artificial intelligence. And (a fourth theme) ideas from AI also fed back into the computational study of language with the emergence of the PROLOG programming language. This introduced the idea of 'programming with logic' (see [18]) and gave rise to new 'declarative' approaches to syntax throughout the 1980s.

The previous two paragraphs are highly impressionistic, but they point towards something important: somewhere in the late 1960s and early 1970s such fields as linguistics and computer science became intertwined, subfields such as computational linguistics and AI crystallised, and technical and philosophical logic played an important role in these developments. What I have written is not serious history — writing a detailed account of these developments would be a daunting task. Nonetheless, I hope what I have said conveys something of the swirl and interplay of research from all these fields, an interplay which was mediated by logic, and which transformed the way we understand logic in the process.

In the space of this paper, this is about as far as I can go to explain what the New

Trivium is and where it came from. There is a great deal more that I could add. For example, more recent work on the interplay between logic and cognitive psychology (the key text here is Lambalgen and Stenning [27]) is also relevant. But the basic point is (I hope) clear. In some sense, the study of logic, language, interaction and cognition become mathematised. We might say: the quadrivium invaded the trivium. But the mathematics at the heart of this invasion was logic.

I believe that we now have a New Trivium on our hands — and it's a *lot* deeper than the medieval one. But it's also, quite simply, just plain *bigger*: there is an awful lot to learn and thus an awful lot to teach too. And as my goal is to find ways of bringing it to humanities students with little, if any, background in logic, the word 'quixotic' may seem fully justified.

4   A First Attempt

In my first semester at RUC I taught a brief (six lecture) course for Masters students. The aim was to give them a taste of work related to my own research. I knew that most of them had little or no background in logic, so I decided to teach a course on natural language semantics that emphasised "natural language metaphysics". This seemed an ideal choice: it let me introduce New Trivium topics, enabled me illustrate how modern (semantically driven) approaches to inference could illuminate linguistic and philosophical issues, and — last but not least! — it let me sneak in a little logic through the back door.

The phrase "natural language metaphysics" is due to the late Emmon Bach (see, for example, Bach [2, 1]). Bach emphasised the role of ontological modelling in natural language semantics: what must the world be like to make language (and in particular, certain patterns of inference) work as they do? He put "natural language" in front of "metaphysics" to emphasise his neutrality as to whether the enterprise counted as 'real' metaphysics.[7]

I have always liked his phrase, because it points towards a key issue: the interplay of logic and ontology. Moreover, many of the neatest illustrations involve the way we use language to talk about time, an appealing topic, and one loaded with examples. Furthermore, I had successfully taught similar introductory courses at ESSLLI, NASSLI and Linguistics Institute summer schools. So I assigned accessible (and non-technical) reading such as selections from Reichenbach [25] and Prior [24] on tense, Vendler on verb classification [30] and Davidson [6] on events — and off we went.

---

[7]Bach also made an interesting attempt to teach formal semantics informally: see Bach [3], which is based on a lecture series he gave at Tianjin Normal University, China, in 1984.

Davidson's article provides a good illustration of the sort of ideas I wanted to convey: it shows why abstract entities can be important for inference. To make the issue concrete, ask yourselves how you would use first-order logic to represent the following sentence:

John ate.

The answer may seem obvious: surely we can just represent it using the simple atomic formula *ate(john)*? Well, no. Because if we choose this representation it is not clear how to represent the following (closely related) sentences:

John ate a big kahuna burger.
John ate a big kahuna burger in the kitchen.
John ate a big kahuna burger in the kitchen at 10.00 o'clock.
John messily ate a big kahuna burger in the kitchen at 10.00 o'clock.

If we insist on representing "ate" by a one-place predicate symbol we face the problem that in the second example it expresses a two-place relation (between John and a burger), in the third example a three-place relation (between John, a burger, and a location) and in the fourth example a four-place relation (between John, a burger, a location and a time). The fifth example complicates matters more by asserting something about the manner in which John ate, and by adding further adverbial modifiers we can drive the arity as high as we like.

These sentences are logically interrelated. In particular, the longer sentences entail the shorter ones, and indeed entail many other variant sentences not listed. For example, we have that:

John messily ate a big kahuna burger in the kitchen at 10.00 o'clock
$$\models$$
John ate at 10.00 o'clock.

However simple-minded representational strategies (for example, using several "ate" predicates of different arity) make it difficult to mirror these logical relationships in first-order logic. And this leads to Davidson's point: we should *reify*. Intuitively, the above sentences are talking about an event. So we should regard the verb "ate" as introducing an abstract entity — an event — and work with first-order representations which let quantify over them. If we do this, our representational difficulties vanish: once events are 'things', we can simply attribute properties to them, and let them participate in relations. Moreover, adopting this strategy means that the required inferential relationships hold automatically. For example, the previous natural language entailment is transparently mirrored by the following first-order entailment:[8]

$$\exists e(eating(e) \land manner(e, messily) \land agent(e, John) \land patient(e, BKB)$$
$$\land\ location(e, kitchen) \land time(e, 10.00))$$
$$\models$$
$$\exists e(eating(e) \land agent(e, john) \land time(e, 10.00))$$

Davidson's work shows the interdependence of logic and ontology. This example is fairly typical of the sort of high-level issue I discussed in the course. I also like it because it shows the neutrality — indeed, the emptiness — of logical notation. First-order syntax can be applied in many ways to one and the same semantical problem, and some of these may be very bad (multiple predicate symbols for "ate") and others much better (letting variables range over events). But the notation is merely a useful tool, it's the way we conceive the world that determines success or failure of our modelling.

Nonetheless, I was dissatisfied with the course: I failed to convey the sense of intellectual excitement that drives research in natural language semantics. The students were politely interested in what I had to say, but little more. Why?

I don't think it was because I made use of logical notation. In my experience, if logical formalism is casually introduced, aided perhaps by a few brief words of explanation, students tend to pick it up without complaint.[9] I did notice that the students were unused to thinking recursively — the idea of building new structures (such as sentences) by repeatedly gluing old ones together was clearly novel. But this did not seem to be the source of any obvious difficulties.

The problem, essentially, was motivation. The fact that the course was about language was (initially) interesting — but it examined language in more detail and at a lower level than they were used too. Language is a car they wanted to drive; seeing me poke around under its hood, discussing how the various pieces fitted

---

[8] For readers who have not encountered natural language semantics before, a small caveat: events are not unproblematic, there are many different ways of modelling them, and there are interesting ways of avoiding their use altogether. But these issues are not relevant to this paper.

[9] Some logicians are happy letting beginners 'pick up' logical notation and proof-theoretic skills in this way, others hate the idea. I think it depends on the students, and on what the goal of the course is — but that said, to overemphasise rigour can be a *serious* pedagogical blunder. Rigour (at least as conventionally conceived) plays little role in the way many core mathematical skills are taught. At school, important calculus skills such as integration by parts and integration by substitution are taught by example: students are expected to 'grasp' the basic ideas from the examples given and to apply them to new problems. Formal definitions play little or no role in the process, and this is *not* a sign of anarchy: it's an effective way of 'learning to talk calculus'. Letting beginners 'learn to talk logic' in a similar way is often a sensible option.

together, didn't fire their interest in the way I hoped. They became, in essence, passive spectators. Some students appreciated the interplay of logic, language and ontology — but they lacked the background to develop the ideas further or to use it in their own work. If I wanted to get students interested in logic, I needed another approach.

5 Problematising Logic

Which brings us back to the critical thinking course. This grabbed the students' interest because it (vividly) problematised critical thinking right from start. Problematisation turned students into participants, and gave focus to the course: its goal swiftly became obvious: to find answers to the question *What is critical thinking and why is it so hard?* But can *logic* be problematised in a similar fashion?

First, what do I mean by 'problematising' something? I mean this in two senses. The first is fairly standard. It means, roughly speaking, to call into question leading ideas, accepted notions, and to examine them (in some sense) critically. This first sort of problematisation is evident in my approach to the critical thinking course.

Logic can certainly be problematised in this first sense.[10] Logic is a word that is used and misused, overused and underused. There is a sense that it points to something fundamental, but its status is ambiguous: a solution to a legal or personal problem may be applauded because it is 'logical' — but calling something 'too logical' usually means that some empathic or imaginative ingredient is missing. And many other (often contradictory, or near contradictory) notions swirl around the word: logic is something fixed, there is only one true logic, logic is good for some things but not for others, there are many logics, there is only one logic, logic is neutral, we are born logical, we are not born logical, and so on. Clearly there is much here that humanities students in general (and philosophy students in particular) should find important. But is there a way of using these questions to drive a course on logic? And in a way that leads to the New Trivium? This brings me to the second sense of 'problematise'.

This sense is more playful, but is at least as important as the first. It means, wherever possible, to present key ideas via problems, or puzzles that students can solve (or fail to solve) both singly and in groups. Nor is 'solution' really the point. The problems I have in mind are often more like stories to reflect upon. For the role

---

[10]Note that problematising logic seems to mean making some sort of logical critique of logic. Isn't this inherently circular, dangerous, perhaps even paradoxical? Well, yes! But (as the history of logic demonstrates) this is territory that logicians love to explore and have explored with vigour, imagination, and insight. Thanks to their efforts, this *terra* is no longer *incognita* — and is certainly no terror either.

of these problems, puzzles, stories is to make the underlying issues *vivid*. Certain puzzles are useful tools for teaching logic, because they call into question conceptions and misconceptions about logic in ways that are hard to forget. I view such puzzles as building blocks in the quest to teach logic without teaching logic.

To illustrate this I will discuss two problems that I have experimented with in both lecture and project-based settings. Both problems are drawn from the psychological literature. The first is the Wason Four Card Problem (also known as the Wason Selection Task) which played an important role in the first critical thinking lecture. The second is the Sally-Anne task, a classic false-belief task from the Theory of Mind literature. Both problematise certain aspects of logic in the first sense: that is, they call into question, or throw under a novel light, certain key assumptions of logic. But both problematise in the second sense too: the Wason Four Card Problem simply *is* a little puzzle and the (two versions of) the Sally-Anne task I shall discuss lead to surprises. Both can be hijacked for pedagogical purposes.

Let's begin with the Wason Four Card Problem (see [31]). In this problem, four cards are displayed. Each card has a colour on one side, and a number on the other. Suppose we place four cards on a table so that *3*, *4*, *Red*, and *Brown* are face up. Which cards must be turned over to verify whether the following generalization holds: *If there is an even number on one side, then the colour red is on the other*?

Now, if we turn this into propositional logic, we (as logicians) see that the crucial cards are the *4* and the *Brown*. We need to turn over the *4* in order to verify that the other side is *Red*. We also need to turn over the *Brown*, as doing so could falsify the stated generalisation. It's also clear that the other two cards (the *3* and the *Red*) are irrelevant to the task at hand. We could sum up the situation by saying: turning the relevant cards over corresponds to applying the inference rules Modus Ponens and Modus Tollens, while turning the irrelevant cards overcorresponds to using the fallacies of Denying the Antecedent and Affirming the Consequent.

But to say this is to assume that it is obviously correct to model the situation using classical propositional logic. This flies in the face of the fact that in Wason's original trial fewer than 10% of his subjects successfully solved the puzzle.[11] Moreover, it misses the *pedagogically* important point. The difficulty of the task reinforces what logic teachers know firsthand: a fundamental concept of propositional logic, namely material implication, is abstract and difficult. Teaching material implication can be a tough task: grasping the concept seems to demand a certain suspension of disbelief on the part of the student. What does it mean? Why is its truth table so odd? It certainly doesn't match the meaning of most conditionals in natural language. And this feeling of oddity is entirely natural: the truth-conditional approach

---

[11]I have never had more than a 15% success rate when using the test as a classroom puzzle.

to logic is an abstract invention that only became logical orthodoxy in the 20th century. As logicians we are used to it, which blinds us to its peculiarities. But when teaching it — and especially when teaching it to humanities students with little or no mathematical background or experience of abstract proofs — we should be aware that it's deeply weird. Which brings us back to problematisation (in the first sense): it is an interesting strategy to embrace these difficulties, and let students battle towards understanding this (important but counterintuitive) interpretation of conditionality for themselves.

The Wason task is an ideal playground for novices to do this. And it's a playground that works well with groups — when students are encouraged to work out together what to do, they often reach the 'correct' idea more rapidly. Moreover, thinking about this task opens the door to other logics, and the spirit of the New Trivium encourages thinking in terms of *logics* (plural) rather tham *logic* (singular). For it's *not* obvious that one *should* model the Wason Task in classical propositional logic. I can't explore this point here, but will refer the reader to Lambalgen and Stenning [27] which discusses the task in terms of *deontic* (as opposed to declarative) conditionals. But the key pedagogical point is this: grappling with the task gives students a memorable way of exploring concepts of conditionality. The Wason task was designed to show how badly humans perform 'logical reasoning'. This makes it a remarkably effective tool for teaching logic, one that lets students gain real insight into what 'logical reasoning' is all about.

Let's turn to the Sally-Anne task. This is a story to reflect on rather than a puzzle to be solved and I'll present two versions of it. Here's the basic scenario.

An experimenter, using two dolls called Sally and Anne and various other props (a basket, a box, and a ball) is telling a story to a child — the dolls and props are used to act out the story, making it more vivid. The story the experimenter is telling runs something like this:[12]

> *This is Sally, and this is Anne. And this is Sally's basket, and that's Anne's box. Look: Sally is putting the ball into her basket! And now she's going away. Bye bye Sally! But what's Anne doing? She's taking the ball out of the basket and putting it in her box. But Sally doesn't see it because she went away. Tricky Anne! Oh, look. Now Sally's coming back in again.*

The experimenter then asks the child the key question:

---

[12]On YouTube you can usually find videos of psychologists administering the Sally-Anne task (and other first-order false-belief tasks, such as the Smarties task) to children.

Where will Sally look for her ball?

Now for the first surprise. Most three-year-olds answer: "In the box". On the other hand, almost all five-year-olds will give the correct answer: "In the basket". Somewhere around the age of four, children come to realise that others can have beliefs about the world that can be *false*.[13]

And there's a second surprise: children with Autism Spectrum Disorder (ASD) usually fail first-order false-belief tasks until a significantly later age; only about 20% pass at the same age as typically developing children. This was first shown by Baron-Cohen, Leslie and Frith in a seminal study (see [5]), and Baron-Cohen later argued that ASD involves a kind of 'mindblindness', and even used this as the title for a book (see Baron-Cohen [4]).

But let's move on to *second-order* false-beliefs. In the version of the Sally-Anne task just given, the child was asked to judge Sally's belief about where the ball is. This is a first-order belief: it is about Sally's belief about the world. But beliefs can also be about other beliefs, and such beliefs are called second-order beliefs. We can easily adjust the first-order Sally-Anne task so that it becomes a false-belief task for second-order beliefs. The new story is the same as the previous version — until the part where Sally leaves the room and Anne puts the ball in the box. We then continue as follows:

Oh, look. Sally didn't really go away! She's peeking through the window! She sees Anne move the ball! But Anne doesn't notice. She doesn't see that Sally is watching. Now Sally's coming back in again.

The experimenter then asks the child the key second-order question:

Where does Anne think that Sally will look for her ball?

The second-order nature of the question is clear from its syntax: we have transformed the first-order question *Where will Sally look for her ball?* into

*Where will **Anne think that** Sally will look for her ball?*

which has one level of sentential embedding more.

---

[13]The original experimental work on first-order false-beliefs was carried out by Wimmer and Perner's in 1983 (see [33]). The basic findings have been repeatedly replicated under a wide range of conditions: the effects of language, culture, story complexity, verbal versus non-verbal presentations, and much else besides, have been investigated. The basic findings still stand.

Once again, surprises are in store. Typically developing children learn to handle second-order false-beliefs somewhere around the age of six. Moreover, second-order false-beliefs seem to be extremely difficult for children with ASD.[14]

When students hear that three-year-olds fail the first-order task, they are usually surprised, even sceptical. Something unexpected — something *strange* — is going on. What exactly?

A lot. We are usually so comfortable with our everyday habit of reasoning about the beliefs, desires and intentions of others that it never occurs to us that this (on reflection, quite remarkable) capacity might have had to be learned. We all realise that we learnt *language* at some stage — but, when we were three, did we really not understand that *beliefs* held by others could be false? It seems we did not.

Moreover, not only does there seem to be a conceptual 'jump' at the age of four (when typically developing children learn that beliefs about the world can be false) but there seems to be a similar 'jump' at the age of about six when typically developing children learn to reason about beliefs about beliefs. Indeed we might say: this is when typically developing children reify the belief concept. Or at least: this is when they begin to act as though the world contains abstract entities (beliefs) that can not only be about concrete entities (like cabbages and kings) but also about beliefs themselves. We are close to the territory I tried to teach using Davidson on events: the interdependence of logic and ontology, and why reification and abstract entities are important. But approaching these ideas via the Sally-Anne tasks has proved more pedagogically compelling for my students.

There are a number of other puzzles I use as pedagogical tools, drawn from various disciplines. One is the well-known Muddy Children puzzle (see [8]). This is useful because it gives students insight into recursive thinking — and works beautifully in group settings. Moreover, some classic philosophical material, such as the Master Argument of Diodorus Cronus (see [11] for an excellent presentation of this, and related puzzles) also lends itself to this style of teaching. I am slowly assembling a collection of building bricks (and baking a few of my own).

6 Can it all be put together?

I have indicated what the New Trivium is, described a disappointing attempt to approach it using very little logic, and zeroed in on what I think is the crucial pedagogical idea: problematising logic. But while problematisation worked in the

---

[14]The original experimental work on second-order false-beliefs was carried out by Perner and Wimmer in 1985 (see [23]). But far less is known about them. For a book-length treatment of the subject, see Miller's *Theory of mind: Beyond the preschool years* [20].

critical thinking course, I have as yet little evidence that it can provide the framework for an entire logic course. For until now I have only used it to provide logical interventions in other courses. For example, I now routinely use the Wason Selection Task when I need to discuss the basics of propositional logic in (say) Epistemology or Philosophy of Science classes, and I plug in other problems as and when they seem appropriate in other courses.

Such interventions can be useful. For example, students seem to get more out of Miranda Fricker's [12] analysis of the trial scene in Harper Lee's *To Kill a Mockingbird* after they have worked though the Sally-Anne and Muddy Children scenarios. These scenarios sensitise them to the subtleties involved in reasoning about knowledge and belief, a useful frame of mind for approaching Fricker's work.

But can problematising logic provide the conceptual skeleton for an entire logic course? Can we *really* teach logic without teaching logic? And can we present the New Trivium in this way? At this point, however, all I can offer is speculation (probably coloured by wishful thinking). To keep the remaining discussion brief and focussed, I'll present it question-and-answer style.

Q. *Do you really see useful course(s) emerging out of such problems?*
A. Yes, I think so. The key will be to find the right examples, and the right sequence for them. And the goals of the course will have to be crystal clear to get this right.

Q. *You're essentially saying you'll need to impose some kind of overall architecture. But this means that you are guiding the students. Doesn't that undercut your claim that this is about "teaching logic without teaching logic"?*
A. Well, yes. But I used the phrase to dramatise a key idea: enabling students to unearth crucial concepts themselves.[15]

Q. *I can see that this approach might be useful to stimulate philosophical discussion, but is it an effective way to teach concrete skills?*
A. I think so. It would be easy to motivate (say) the tableaux method in such a course.[16] Or natural deduction. Or resolution. Or how to systematically translate natural language expressions into logic.[17] There are many useful skills which fit well with the approach.

---

[15]There is a long tradition of trying to do something similar in mathematics. Jumping back to antiquity, in *Meno*, Plato tries to establish that learning is recollection by having Socrates elicit a proof of a geometrical theorem from an untrained slave boy. More recently, several mathematics departments have experimented with the Moore method (named after Robert Lee Moore) to 'teach' such subjects as number theory, point set topology, and mathematical logic. The key idea of the method is that the students should prove the key theorems in these subjects themselves.

[16]Wilfred Hodges does something like this in his layperson's guide to logic (see [15]).

[17]See Montague and Kalish [21] for classic exposition. The experience of writing this textbook was one of the motivations that led Richard Montague to develop Montague grammar.

Q. *But won't this always be second-best to a more rigorous presentation?*

A. I don't believe that at all. It's true that when it comes to technicalities, and confidence with symbol manipulation, mathematicians and computer scientists usually have an initial advantage. Moreover, linguists, because of their familiarity with phrase structure trees and other modes of recursive representation, are likely to have an advantage over other humanists. But a course of the kind just described could convey a lot of these skills. If you've thought hard about the Muddy Children, and have done your fair share of natural deduction proofs, you'll soon be ready to move on to the next stage.

Q. *So there is a next stage? What is it?*

A. Start attending ESSLLIs! Actually, that's not a joke. A good measure of the success of such a course would be that it prepared humanities students to attend an ESSLLI or a NASSLLI or something similar and get something useful out of it.

Q. *OK, but if this is such a great idea, why not use it for everyone? Why all this talk about humanities students?*

A. The point is this: mathematics and computer scientists are (pretty much automatically) equipped for the study of logic, at least as far as technical prerequisites are concerned. Linguists too — indeed linguists not only have some of the skills required for logic (recursive thinking, for example) and they also have some motivation to study it (for example, to let them get to grips with formal semantics). Linguistics students would probably have enjoyed my course on natural language metaphysics way more than my philosophy students did.

Motivation is always an issue – and problematisation offers a compelling handle here. You're only going to get one chance to get this stuff across, so you've got to work in a way that *simultaneously* provides strong motivation and provides technical and intellectual tools. Humanities students need both. And problematisation (in both senses) seems a promising way of providing what they need.

Q. *Hmmm . . . isn't this paper essentially a long daydream about a new textbook?*

A. Maybe. Or perhaps it's the worlds longest ESSLLI course proposal. Teaching logic without teaching logic need not be paradoxical, whimsical or even Quixotic. But it will involve a lot of hard graft.

7 Concluding Remarks

Arguably, something worth calling the New Trivium has emerged over the last thirty or so years. In the late 19th and early 20th century logic entered a revolutionary phase: its mathematical content became visible and (perhaps more importantly) it entered into a symbiotic relationship with computer science, a field it helped to create. But while old logical themes (such as links with language, cognition, and knowledge) were temporally overshadowed, they re-emerged in the late 1960s and become relevant again, giving rise to what I have called the New Trivium.

But this paper has focussed not so much on the New Trivium as on how to teach it, and in particular, on how to teach it to humanities students with little or no knowledge of logic. Much of the paper argued that problematising logic was a promising way of teaching it without teaching it. But there is a further question I did not address: why bother with teaching it to this audience at all? Why not just accept that logic and language (in cahoots with information) have flown the coop? The mathematical quadrivium has swallowed the linguistic trivium and that is pretty much that. The humanities are dead if not yet buried.

Because I don't believe it's true. Many humanists insist that what makes their work different from (say) that of natural scientists is its interpretational nature. To put it another way: what they do is different because it fundamentally involves language-oriented skills, the sort of skills rooted in the original trivium. But the recursive, self-referential nature of modern logic and computing, aligns well with the endless hermeneutic cycling of the humanities. The New Trivium, rooted in the logics of the late 20th century has something to say about how meanings are made, how they are expressed, how they multiply and how they lead to information flow — and the humanities may well have the most important things to say on all these topics. It is time for a double reinsertion: logic into the humanities, and humanities into logic. And it is ultimately to make this reinsertion possible that I think it is worthwhile to learn how to teach logic (so to speak) without teaching logic. For, with luck, some students will see how to live in both worlds at once — and such students will be the ones who will work out what needs to be done next.

References

[1] Emmon Bach. On time, tense, and aspect: An essay in English metaphysics. In Peter Cole, editor, *Radical Pragmatics*, pages 63–81. Academic Press, 1981.

[2] Emmon Bach. Natural language metaphysics. In R. Marcus, G. Dorn, and P. Weingartner, editors, *Logic, Methodology and Philosophy of Science*, volume 7, pages 573–595. Amsterdam, North Holland: Elsevier Science Publishers BV, 1986.

[3] Emmon Bach. *Informal lectures on formal semantics*. State University of New York Press, 1989.

[4] Simon Baron-Cohen. *Mindblindness: An essay on autism and theory of mind*. MIT press, 1997.

[5] Simon Baron-Cohen, Alan Leslie, and Uta Frith. Does the autistic child have a theory of mind? *Cognition*, 21(1):37–46, 1985.

[6] Donald Davidson. The logical form of action sentences. In Nicholas Rescher, editor, *The Logic of Decision and Action*. University of Pittsburgh Press, 1967.

[7] John Dewey. *How we think*. Courier Corporation, 1997. First published in 1909.

[8] Ronald Fagin, Yoram Moses, Joseph Halpern, and Moshe Vardi. *Reasoning about knowledge*. MIT press, 2003.

[9] Cordelia Fine. *A mind of its own: How your brain distorts and deceives*. WW Norton & Company, 2008.

[10] Alec Fisher. *The logic of real arguments*. Cambridge University Press, 2004.

[11] Melvin Fitting and Richard Mendelsohn. *First-order modal logic*. Springer Science & Business Media, 2012.

[12] Miranda Fricker. *Epistemic injustice: Power and the ethics of knowing*. Oxford University Press Oxford, 2007.

[13] Paul Grice. *Studies in the Way of Words*. Harvard University Press, 1991.

[14] Jaakko Hintikka. *Knowledge and belief: an introduction to the logic of the two notions*. Cornell University Press Ithaca, 1962.

[15] Wilfrid Hodges. *Logic*. Penguin, 2001.

[16] Daniel Kahneman. *Thinking, fast and slow*. Macmillan, 2011.

[17] Hans Kamp and Uwe Reyle. *From discourse to logic: Introduction to modeltheoretic semantics of natural language, formal logic and discourse representation theory*. Springer Science & Business Media, 1993.

[18] Robert Kowalski. *Logic for problem solving*. Ediciones Díaz de Santos, 1979.

[19] John McCarthy and Patrick Hayes. Some philosophical problems from the standpoint of artificial intelligence. *Readings in artificial intelligence*, pages 431–450, 1969.

[20] Scott Miller. *Theory of mind: Beyond the preschool years*. Psychology Press, 2012.

[21] Richard Montague, Donald Kalish, and Gary Mar. *Logic: Techniques of Formal Reasoning*. Harcourt, Brace, Jovanovich, New York, 1980.

[22] Richard Montague and Richmond H Thomason. *Formal Philosophy: Selected Papers of Richard Montague; Ed. and with an Introduction by Richmond Thomason*. Yale University Press, 1976.

[23] Josef Perner and Heinz Wimmer. John thinks that Mary thinks that... attribution of second-order beliefs by 5-to 10-year-old children. *Journal of experimental child psychology*, 39(3):437–471, 1985.

[24] Arthur Prior. *Past, present and future*. Clarendon Press Oxford, 1967.

[25] Hans Reichenbach. *Elements of symbolic logic*. The Free Press, 1947.

[26] Paul Ricoeur. *The rule of metaphor: The creation of meaning in language*. Psychology Press, 2003.

[27] Keith Stenning and Michiel Van Lambalgen. *Human reasoning and cognitive science*. MIT Press, 2008.

[28] Stephen Toulmin. *The uses of argument*. Cambridge University Press, 2003. First published in 1958.

[29] Hans van Ditmarsch, Wiebe van der Hoek, and Barteld Pieter Kooi. *Dynamic epistemic logic*, volume 337. Springer Science & Business Media, 2007.

[30] Zeno Vendler. *Linguistics in philosophy*. Cornell University Press, 1967.

[31] Peter Wason. Reasoning about a rule. *The Quarterly journal of experimental psychology*, 20(3):273–281, 1968.

[32] Daniel Willingham. Critical thinking: Why is it so hard to teach? *Arts Education Policy Review*, 109(4):21–32, 2008.

[33] Heinz Wimmer and Josef Perner. Beliefs about beliefs: Representation and constraining function of wrong beliefs in young children's understanding of deception. *Cognition*, 13(1):103–128, 1983.

Received 11 October 2016

Logic for Fun: an online tool for logical modelling

John Slaney
The Australian National University

Abstract

This report describes the development and use of an online teaching tool giving students exercises in logical modelling, or *formalisation* as it is called in the older literature. The original version of the site, 'Logic for Fun', dates from 2001, though it was little used except by small groups of students at the Australian National University. It is currently in the process of being replaced by a new version, free to all Internet users, intended to be promoted widely as a useful addition to both online and traditional logic courses.

Keywords: Logic Teaching, Online Learning, Logical Modelling, First Order Logic.

1 Background: Formalisation

In introducing undergraduates to formal logic, we attempt to impart a range of skills. In a typical "Logic 101" course, the most prominent of these involve manipulation of calculi: devising proofs, usually using some form of natural deduction, constructing semantic tableaux or the like. We also ask students to formalise natural language sentences—often specially constructed to involve awkward nesting of connectives or strings of quantifiers—and may hope that they acquire some facility in critical reasoning and perhaps an appreciation of some wider issues connected to logic, be they mathematical, computational, philosophical or historical. Some of these things we teach better than others. Although devising proofs is traditionally a stumbling block, most students do in fact become tolerably adept at handling the technical details of natural deduction. Where we fail is rather in teaching them to say what they mean in the abstract notation of logic: many students who can mechanically

A short version of this paper was presented in the conference *Tools for Teaching Logic* in Rennes in June 2015. Thanks are due to the participants in that conference for illuminating conversations and ideas.

construct a proof remain depressingly unable to write a well-formed formula to express even a simple claim about a domain of discourse.[1] Barwise and Etchemendy, for instance, comment:

> The real problem, as we see it, is a failure on the part of logicians to find a simple way to explain the relationship between meaning and the laws of logic. In particular, we do not succeed in conveying to students what sentences in FOL mean, or in conveying how the meanings of sentences govern which methods of inference are valid and which are not. [3], p.13

We should find this situation alarming. Mechanical symbol-pushing for the purposes of simple proofs is easy to teach, tolerably easy to learn—at least, if students can be brought to work at it—and almost useless once the course is finished. On the other hand, the ability to read and write in the notation of formal logic, to use this as a medium for knowledge representation, to analyse and to disambiguate, is the most important skill students can take away from an introductory logic course, and it is a skill most of us can claim less success in teaching.

The reasons why students find formalisation so hard are not difficult to discern. The problem of rendering a description into formal notation has no unique solution and there is no easy way to know whether a putative solution is right or wrong. There is also no simple algorithm for treating such problems, so "surface" learning is ineffective. The ony successful method, in fact, is to understand both the natural and formal languages and to match the two. Faced with this challenge, students not infrequently give up.

It was against this background that the tool *Logic for Fun* [9, 10] was devised around 15 years ago. *Logic for Fun* is a website on which users are invited to express a range of logical problems and puzzles in such a way that a black-box solver on the site can produce solutions. It was never tied to a prescribed course, but was intended to be used as an adjunct to undergraduate courses, whether those be in logic, critical reasoning, artificial intelligence or other fields, or by interested individuals outside the context of formal instruction. The language in which problems are to be expressed is that of a many-sorted first order logic, extended slightly with a few built-in expressions and modest support for integers. The solver takes as input a set of formulae in this language and searches for finite models of this set. If it finds a unique model, this is almost certainly a solution to the problem. More usually it

---

[1]Evidence for this claim is anecdotal, but strong. My own appreciation of it was sharpened in 2011, when analysis of results from a class of 61 students showed grades on proof construction that were on average above their grades for other courses, but after 13 weeks of study more than a third of them were unable to express 'The bigger the burger the better the burger' adequately in the notation of first order logic.

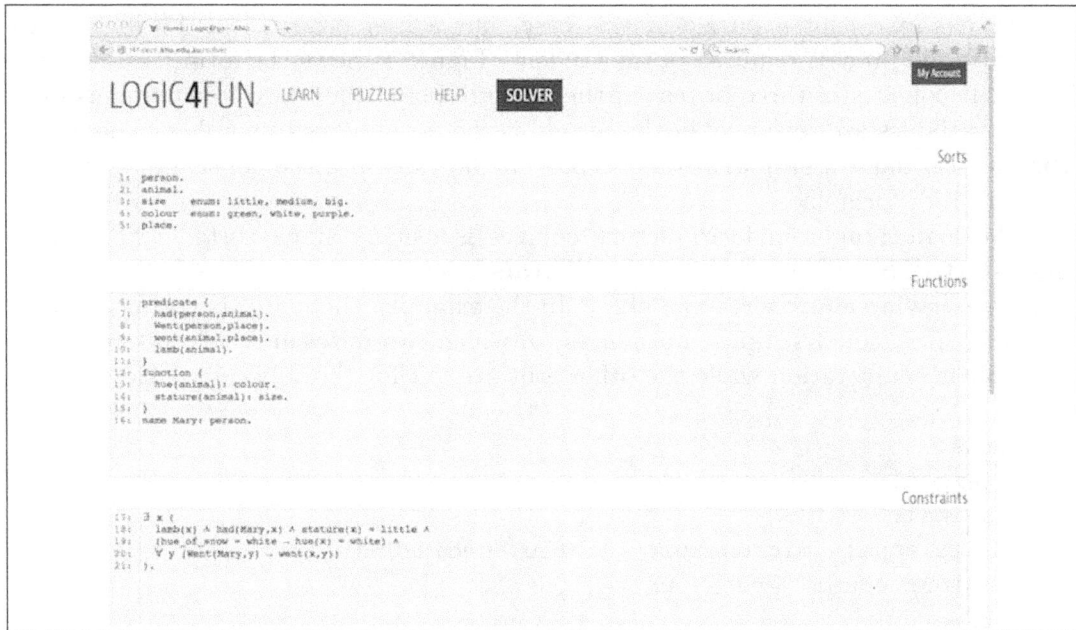

Figure 1: Screenshot of the window for problem input

either reports syntax errors in the encoding or else says that there is no solution. The user (i.e. the student) then debugs the encoding until it is correct. This has several advantages over traditional formalisation exercises:

1. The sentences to be formalised constitute a meaningful problem, rather than looking like isolated examples of things it may be tricky to express formally;

2. There is an easily graspable concept of a *correct* solution, so students know whether they are right or wrong;

3. Feedback is always accurate and *immediate*, rather than coming a week after the homework was handed in;

4. Because there is no feedback latency, because there is a goal (solving the problem) and because the machine is infinitely patient, students will put time and effort into their work, to a degree never seen in the traditional setting;

5. For the problem encoding to count as correct, it must be completely accurate, as the computer will not accept hand-waving, so the value of rigorous attention to detail is constantly emphasised.

An example (not a puzzle in this case, but a first order theory) will help to illustrate the process required of the student. Figure 1 shows the form used for text input. It consists of three boxes: in the first are listed the "sorts" or domains over which variables are to range; in the second is the vocabulary of non-logical symbols (predicates, names, etc) with their types; in the third are the constraints, written as first order formulae.

The logic taught in introductory courses is usually single-sorted rather than many-sorted, but the experience over 15 years has been that students find it easy to adjust to having many sorts available.[2] In the example, there are six sorts (persons, animals, sizes, colours, times and places) of which two (sizes and colours) are given by explicit enumeration while the other four are left for the solver to decide.

```
Sorts:
  person.
  animal.
  size enum:  microscopic, little, medium, big.
  colour enum:  green, white, purple.
  time.
  place.
```

The domains of these sorts are disjoint. We wish to say that Mary had a little lamb, so as vocabulary we declare 'Mary' as a name, 'had' as a relation between persons and animals, 'stature' as a function of animals and 'isLamb' as a predicate:

```
Vocabulary:
  predicate {
    had(person,animal).
    isLamb(animal).
  }
  function {
    hue(animal):  colour.
    stature(animal):  size.
    location_p(person,time):  place.
    location_a(animal,time):  place.
  }
  name Mary:  person.
```

---

[2]Single-sorted encoding of problems is perfectly possible in *Logic for Fun*, in case anyone really wants or prefers it, but for most problems it is not recommended.

The convenience of using a many-sorted logic for knowledge representation purposes is immediately apparent. It enables us to specify the types of predicates and function symbols separately from the formulae in which they occur. The hue and stature functions, for example, map animals to their colours and sizes; to model the constraints will be to determine which functions to assign to them. This not only removes the need for sort specifications in the antecedents of the constraints, but it allows the type checker to detect many errors which might otherwise lead to nonsense.[3]

Since every user-defined predicate and function symbol is strictly typed, and sorts are disjoint, it is not possible to specify a `location` function which can apply to both people and animals. Hence there are two, one for each sort. We have considered relaxing the syntax rules to allow unions of sorts in argument places, but in fact the ramifications of this would cause more complexity for the user than would be justified by the reduction in artificiality of the syntax.

To finish the example, here is the constraint:

```
∃ x (
    had(Mary,x)  ∧
    isLamb(x)  ∧
    stature(x) = little  ∧
    (hue_of_snow = white  → hue(x) = white)  ∧
    ∀ t (location_a(x,t) = location_p(Mary,t))
).
```

Below the constraints box are buttons, not visible in Figure 1, for running the solver. There is one for generating solutions and another for running in a lightweight mode to check syntax. Having written the description, the student clicks "Solve" and the back-end reasoner—essentially a SAT solver adapted to finite domain first order problems—starts searching for solutions. Naturally, it finds models of this little theory with *very* small domains, including the expected interpretation in which there is one person (Mary), one animal (the little white lamb) and one place which neither of them visits. Amusingly, the solver also finds unexpected solutions, for instance in which the lamb is green—but it is still just as white as snow because snow is purple!

---

[3]Other problem representation languages also benefit from this feature. The constraint modelling language Zinc [4, 6] for example has an elaborate type system, whereas its subset Minizinc [7] treats almost everything as a number. In Minizinc, you can add an employee to a day of the week and divide by a truck, and no type error is detected. Many-sorted first order logic is a valuable step towards fully typed languages, with all the advantages they confer.

In the pedagogic context, this provides a good opportunity for the teacher to make some points about interpretations and truth conditions and the semantics of the material conditional. When this kind of situation arises in modelling a puzzle which should admit only one solution, the student must invent more constraints to supply the missing information (e.g. that snow is white and that Mary went to school). In order to do this, they have to think about the semantics of the problem, render it into first order logic and understand the relationship between the formulae they have written and the satisfying formal interpretations.

Of course, some consistent first order formulae have no finite models, and even where finite models do exist, there may be none sufficiently small to be presented. The existence of models and of finite models is undecidable in general. However, this is not a matter of great concern, firstly because the problems on the *Logic for Fun* site are deliberately chosen to have easily discovered small models, and because the solver is in any case set to time out after three seconds (or another short time if the user chooses) so completeness of the search is neither expected nor really desired.

The most significant precursor of *Logic for Fun* was Tarski's World [2, 1] which provides exercises in reading and writing formulae expressing facts about situations. The use of problems rather than simply states of affairs as the basis, and of the user's "freestyle" choice of vocabulary, mark significant differences between *Logic for Fun* and earlier software. More recently, websites have appeared using functionality similar to that of *Logic for Fun* for other purposes. A good example is MiniZinc which has been used to teach aspects of constraint programming—there is even a MOOC based on it [12]—though the emphasis there is partly on efficient encoding and search methods rather than purely on logical modelling.

2 Structure of the exercises

The problems given as exercises on the site are divided into five levels: Beginner, Intermediate, Advanced, Expert and Logician. The boundaries between levels are not really definite, but students like the idea of progressing through levels in the style of a game. "Beginner" problems are fairly trivial to represent and solve, and are designed to get students through the phase of learning to use the site, teaching them how to declare vocabulary, write constraints and read the solver output. Figure 2 shows one of the "Beginner" problems with a suggested encoding. Note that students actually need to learn a good deal about function declarations and other syntax details in order to master this level, but that the logic is not deep. "Intermediate" problems are mainly logic puzzles of the kind found in popular magazines, often calling for bijections between sets of five or six things satisfying a list of clues. There is a long

There are four children, Alice, Boris, Claire and David. Their ages (not in order) are 6, 7, 8 and 9. Each child has either 1, 2, 3 or 4 jellybeans (a different number each).

1. Alice's age plus her beans is the same as Boris's age plus his beans.
2. Claire's age is two more than the number of Alice's beans.
3. Alice's age is equal to the number of beans she and Claire have altogether.

How old is David, and how many beans has he?

Sorts

```
1:  child enum: A, B, C, D.
```

Functions

```
2:  function age(child): int    {all_different hidden}
3:  function beans(child): int {all_different hidden}
4:  name David's_Age: int.
5:  name David's_Beans: int.
```

Constraints

```
 6:  ALL x (
 7:     age(x) > 5 AND age(x) < 10 AND
 8:     beans(x) > 0 AND beans(x) < 5
 9:  ).
10:
11:  age(A) + beans(A) = age(B) + beans(B).
12:  age(C) = beans(A) + 2.
13:  age(A) = beans(A) + beans(C).
14:
15:  David's_Age = age(D).
16:  David's_Beans = beans(D).
```

Solver Output

```
Model 1

David's_Age = 8

David's_Beans = 1
```

Figure 2: A "Beginner" problem, its encoding and its solution

tradition of making these problems gently humorous, and students are usually familiar with the style of problem, so most of them find it fairly easy to get this far with the site. "Advanced" puzzles are not necessarily harder, but have features requiring more sophisticated logical treatment—nested quantifiers and the like. This can cause difficulties for users without a background in logic, so where *Logic for Fun* is used in a logic course, it is worth spending time on several of the "Advanced" problems rather than rushing past them to get to more interesting ones. The "Expert" puzzles are more challenging, and include several state-transition problems from AI planning, for instance. They require students to supply less obvious vocabulary and axiomatisation to represent preconditions, postconditions and frame conditions of actions and so forth. Most of them can be represented in several different ways, using different ontologies and different styles of constraints. This offers opportunities for the teacher to discuss (and for the student to experiment with) non-trivial aspects of formalisation. Finally, the "Logician" section contains problems which hint at applications of logic, for instance to finite combinatorics and to model-based diagnosis.

It is important that all of the suggested problems can be solved quickly by the software behind the site, without requiring coding tricks. This is because the aim is to teach correct logical expression, not constraint programming. Especially for some of the "logician" puzzles, efficiency does matter, as the underlying problems (e.g. minimal hitting set, classical planning, quasigroup completion) are NP-hard, and solver behaviour can be affected by non-obvious things like the order in which functions are declared, but as far as possible the site de-emphasises efficiency and instead lays stress on correctness.

An interesting feature of teaching logical modelling in this way is that concepts are introduced in approximately the opposite order from that in the parallel lectures. The standard structure of a typical logic course is to work from the more abstract levels down to the more detailed ones. We start with the generic idea of inference, then proceed to examine propositional connectives, then move to first order logic with names, predicates and quantifier-variable notation. Then we introduce identity as a special relation symbol and go on finally to deal with function symbols and general terms. We do not usually get as far as many-sorted logic. There are good pedagogic reasons for this order of exposition, as the more intricate parts of logic presuppose the simpler ones, and the details make more sense within a clear framework than they do in isolation. It is undeniably easier to learn to manage boolean operations on sentences than quantification over arbitrary domains, so we aim to give our students facility in manipulating the former before expecting them to tackle the latter.

Logical modelling, by contrast, *starts* with sorts, equations, names and functions.

The very first example in the user guide to *Logic for Fun* is

Find a number x such that $2 + x = 4$

which of course students without a logical background find entirely trivial. Considered as a first order formula, however, it involves both interpreted and uninterpreted individual constants, a binary function symbol with a built-in interpretation, equality as a special binary relation, and integers as a sort—quite advanced material for Logic 101. The key concept right at the start is that of interpreting a formula by assigning values to its uninterpreted symbols. Generality needs to be introduced next. The universal quantifier is used much more than the existential one: at this early stage, existentials can mostly be replaced by Skolem functions (mostly constants) without making students explicitly aware of this substitution. Only after that do we meet the basic connectives. Introductory logic textbooks, though they vary greatly in emphasis and style, tend to follow the same overall direction as logic courses.[4] My experience of using *Logic for Fun* as part of a standard introductory logic course is that the reversed order of topics does create a certain degree of difficulty. It requires the lecturer to spend time explicitly pointing out the relationship between the logical modelling exercises and the rest of the course, as the two strands do not converge until the end and may seem disjoint to many of the students. I have not yet experimented with the possible strategy of inverting the entire course, making logical modelling the centrepiece and working from the detailed and specific towards the more abstract.

3   The logic of *Logic for Fun*

The fragment of logic underlying *Logic for Fun* is chosen to be useful for expressing simple theories over finite domains without departing too far from standard first order logic. Thus the language includes the usual connectives, $\land$, $\lor$, $\neg$, $\rightarrow$, and quantifiers $\forall$ and $\exists$. It also makes heavy use of the identity symbol, $=$, not only to express equations but also to express uniqueness and the like. As noted above, the logic is many-sorted, so that variables are able to range over things of a kind without repetitive antecedents to restrict the constraints, and so that functions can be typed to remove the need for many cumbersome axioms. On interpretation, the sorts correspond to disjoint finite domains. Since the relation of identity makes sense for every domain, the equality symbol is typically ambiguous, though it is a type error to assert an identity between two things of different sorts. The objects of

---

[4]There are exceptions. One of the most notable is the little introduction by Wilfred Hodges [5] used over 30 years ago as a textbook by the Open University and still in print.

a sort may be enumerated explicitly, giving each a canonical name and defining a total order on them, or they may be left unspecified. Upper and lower bounds may optionally be placed on the cardinality of a non-enumerated sort.

Any non-reserved string of characters not containing punctuation or space may be either a name or a variable. It is a variable if bound by a quantifier; otherwise it is a name. It may not be used both as a name and as a variable in the same problem encoding. Names may be declared along with the rest of the vocabulary, or may be used without declaration.

For the purposes of writing formulae, the logical symbols may be written as the English words AND, OR, NOT, IMP, ALL and SOME (all in upper case). This is helpful for students in the early stages of learning to use the site, who may be unfamiliar with standard notation. There is an option to use pure clause form for writing constraints, avoiding explicit connectives and quantifiers altogether in favour of simple punctuation. This was designed to make *Logic for Fun* independent of notation, so that it could be seen as compatible with absolutely any introduction to elementary formal logic. However, pure clause form proved unpopular with students, so this option is disabled in the current version of the site and "normal" first order notation used instead.

There are two built in sorts, int and bool. The latter consists of the two truth values with canonical names FALSE and TRUE (in that order). The former does *not* consist of the integers: since all sorts are finite, and in fact quite small, it consists of the first few natural numbers $0, \ldots, $MAX_int, where MAX_int is set to a fairly low value such as 100. For encoding logical puzzles, it is by default set even lower, at 20 or 30, as the exercise only involves logic, not arithmetic, for which purpose very small numbers are sufficient. Note that identity on the boolean sort is material equivalence, so no special symbol besides '=' is needed for the material biconditional.

More built-in operations are provided, as they have been found useful for expressing finite domain problems. Since each sort is totally ordered in a canonical way—either by explicit enumeration or implicitly—it makes sense to refer to the smallest (first) and largest (last) elements of a sort as MIN and MAX respectively. These symbols may be subscripted with the name of a sort if this is not deducible from the context. The ordering relation on any sort is represented by '<' and '>' as one would expect. Any element may be incremented or decremented by a positive integer, so for instance foo+2 is the item (if any) which comes two after foo in the canonical ordering of its sort. In practice, this is almost always used just to add or subtract one to refer to the successor or predecessor of an object.

Predicate and function symbols, including individual constants (names), may be declared as in the example in Section 1 or in Figure 2. Each has not only a specified type but optionally a list of features. These include the useful feature "hidden", which suppresses printing of the symbol and its value. Binary operators, whether predicates or function symbols, may optionally be written in infix position between their arguments: there is no need to specify this, as the parser accepts both prefix and infix notation, even for the same symbol in the same formula.

Functions of any type may be partial, lacking values for some arguments. By default, functions are total, but may be declared partial when they are specified in the "Vocabulary" box. Certain other properties, such as being injective or "all_different" may be enforced in the same way. Each sort has an existence predicate **EST** (for "there Exists Such a Thing as...") which returns **TRUE** if the expression to which it applies has a value in the domain of the interpretation, and **FALSE** otherwise. This is extremely useful where partial functions are used. In a domain of persons, for instance, we might have a function **spouse** returning an individual's husband or wife. Then (for the artificial purposes of some problem) we may want to say that there are no same-sex marriages within the domain:

```
ALL x (NOT(female(spouse(x)) = female(x))).
```
However, there may be unmarried people, so we want **spouse** to be a partial function, in which case we can write:
```
ALL x (EST(spouse(x)) IMP NOT(female(spouse(x)) = female(x))).
```
to say that if there exists such a thing as x's spouse, then that individual has opposite gender from that of x.

The list of extensions and restrictions of standard first order vocabulary may may seem complicated, but in fact they serve to adapt "pure" first order logic rather modestly for the purpose of easy applicability to finite domain problems.

One question which arose early in the development of *Logic for Fun* was whether to restrict it to first order logic or to allow higher order constructions. Over finite domains, of course, the distinction between first order and higher order expressions is a little artificial, as sets and functions are just more finite objects which could be referenced in a first order way, but there are clear reasons for avoiding them in general for the purposes of this site. Notably, they cause exponential or hyper-exponential increases in the sizes of domains, thus conflicting with the manner in which everything is internally represented to the solver by explicit enumeration, and with the need to solve problems in seconds at worst (more usually in milliseconds). Some second order features, notably allowing reference to the transitive closure of a binary relation, would be useful for knowledge representation in certain cases and have polynomial time propagators, so they could be added without causing an explosion. It is possible that such features may be included at some stage, but as things currently stand they remain unavailable.

By an *interpretation* of the first order language specified for a given problem we mean an assignment of a domain $\mathcal{D}(s)$ to each sort s and an assignment of a

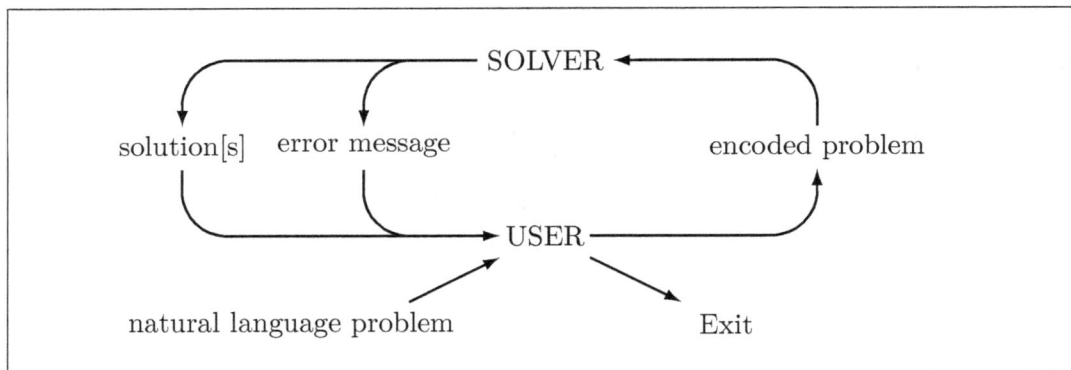

Figure 3: *Logic for Fun* workflow. The main loop is a dialogue between a user and the solver.

function $\mathcal{I}(f)$ of appropriate type to each function symbol f. Each domain is a finite set, and each function symbol of type $s_1, \ldots, s_n \longrightarrow s$ is assigned a function from $\mathcal{D}(s_1) \times \ldots \times \mathcal{D}(s_n)$ to $\mathcal{D}(s)$. A predicate symbol from this perspective is just a function symbol of value type `bool`, and a name is a function symbol as above with $n = 0$. Built in function symbols are given the obvious readings.

As expected, an interpretation is a *model* of the set of constraints specified by the user iff each of them evaluates to `TRUE` in accordance with the standard story about truth for an interpretation. Note that since the language has no free variables, there is no need to treat valuations of variables separately from complete interpretations differing on the values of names. If no model is found by the solver within the time limit, the message "No solution found" is returned. If several models are found, by default up to three are returned. The limit on the number of models may be changed via the "settings" button available to the user. Each model returned is reported by explicitly printing the tables of values of each function for its possible arguments. Students have usually found this style of writing solutions easy to understand.

4 Errors and feedback

The workflow of the site (Figure 3) is one of dialogue between student and machine, whereby the problem in natural language is initially proposed as a challenge to which the student responds by writing formal encodings of all or part of the problem which the solver evaluates. Feedback in the form of error messages or solutions (or lack of solutions) informs the student's next attempt. The cycle is broken when the student decides to terminate or suspend it. Work may be saved at any point for future reference.

Clearly, the educationally effective part of this process is the correction of errors. To put it simply: the tool is only doing useful work when its users are making mistakes. Feedback is therefore the essence of the process. Errors (apart from accidental slips) are fundamentally of two kinds, syntactic or semantic, evoking very different responses from the system. Errors of syntax are caught by the parser or the type checker and reported with explicit messages. For example, if the user writes

```
had(Mary, ∃x(lamb(x))).
```

(presumably trying to say "Mary had some x such that x is a lamb") the solver replies:

```
Input error on line 32:  had(Mary, SOME x lamb(x)).
Type mismatch with argument of had

Detailed diagnostics:  in the formula
    had(Mary,SOME x lamb(x))
the main operator "had" expects argument 2 to be of type animal
but argument 2 is
    SOME x lamb(x)
which is of type bool.

Hints for this kind of error:  check
    (1) parentheses;
    (2) possible typos (e.g.  misspelling);
    (3) how variables and names get their sorts assigned.
```

The suggested possible causes of this kind of error—misplaced parentheses and wrong names—are not guilty in this case, but the key information that there is a boolean formula where a reference to an animal was expected is clearly present. The solver also lists the vocabulary used in the offending formula and writes out the parse tree as far as the parser was able to get before raising an exception. This kind of detail in error messages is an important feature, but such verbosity can become irritating so a possible enhancement for a future version of the site might be to place detailed drill-down under user control.

Users have frequently reported obscure error messages as a major source of frustration in using *Logic for Fun*. Efforts to make error messages more friendly and informative are continuing, as this is an aspect of the site which is perpetually capable of improvement.

Semantic errors are harder to classify and harder to deal with. There are no "canned" solutions written in, so if the encoded problem gets past the parser and type checker, all the solver can do is search for solutions and report what it finds.

Hence the only symptom of misunderstanding on the semantic level is an unexpected solution or (more often) no solution. This is the case whether the error is due to basic misunderstanding of semantics, such as confusion between the conditional and the biconditional, or whether it is a matter of problem representation—using logic correctly to say the wrong thing.

5   Diagnosis tool

The case in which there is no solution is common, and of course the solver's response "No solution found" provides the user with a minimum of information. Techniques for making the feedback more informative include commenting out lines of the encoding and re-running to see whether solutions exist. This can sometimes be effective in pinpointing incorrectly expressed constraints, though it is laborious and the results are not always helpful.

A diagnosis tool designed to help automate the process of isolating incorrect constraints in cases where the encoded problem globally has no solution has been developed: a prototype exists and has been tested by a range of volunteer users, ranging from beginners to experts, but has not yet been incorporated into the live version of the website. As noted, it cannot tell the user what is wrong with their problem encoding, since there is no way of knowing what solution (if any) is desired, but it makes available two types of information:

Approximate models If there is no model of the set of constraints within the parameters set by the sort and vocabulary specifications, the constraints can be marked as "soft" and the solver asked for an assignment of values satisfying as many of them as possible—that is, to solve the problem as a MAX-CSP. The constraints violated by the approximate model are listed. The user may mark some of them as "hard" and re-solve, finding a new model (if there is one) satisfying the hard constraints and approximately modelling the soft ones. Iterating this procedure partially automates the "commenting-out" dialogue, with additional functionality in that optimal approximations rather than arbitrary models are returned.

The back-end solver can run in two modes to search for approximate models. On the site at present, it always searches by depth-first branch and bound, which has the advantges of a complete search method: it stops when the search space is exhausted—often in a fraction of a second—and when it does it returns either a provably optimal approximation or else the information that the hard constraints are unsatisfiable. There is also an option (not currently

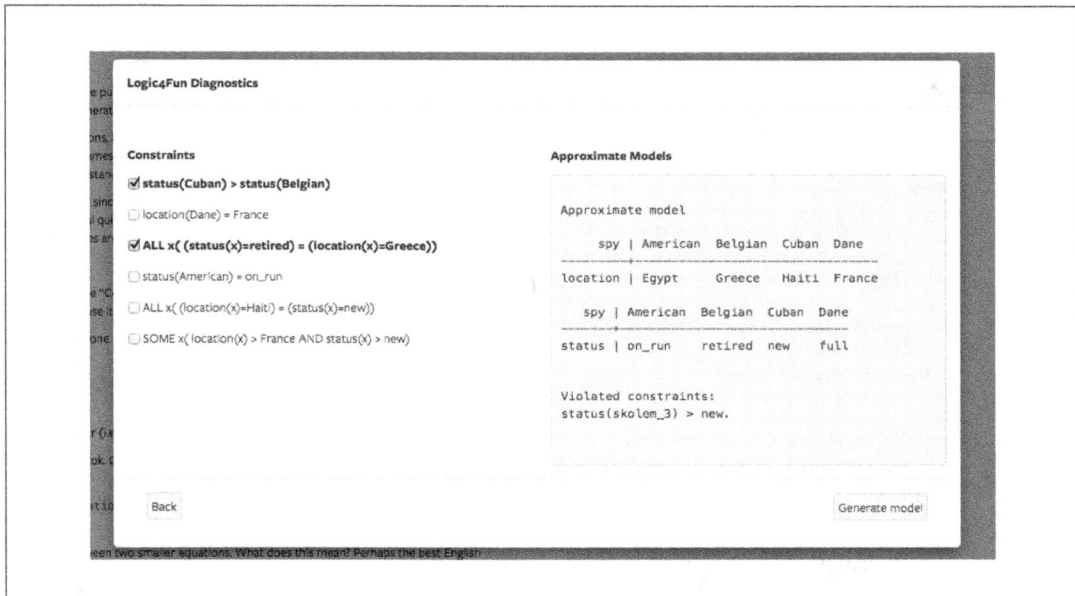

Figure 4: Screenshot of diagosis tool: approximate model

used in *Logic for Fun*) to perform a local search somewhat in the manner of WalkSAT [8]. Since this is an incomplete method, the results it returns come with no guarantee of optimality, and it cannot show unsatisfiability. It does, however, return reasonable models in a reasonable time, even for large or complex problems, so it may have its uses as a fall-back option in cases where the complete search fails. We shall experiment with including it in *Logic for Fun* in future.

Unsatisfiable cores The solver may be asked to identify a minimal subset of the constraints with no solution. Any such subset must contain a contradiction, and so needs correction as at least one of its members is false in the intended model. There may be many unsatisfiable cores in an inconsistent CSP; at present, the diagnosis tool returns an arbitrary one. Finding an optimal (minimal cardinality) unsatisfiable core is computationally difficult: even with an oracle saying whether a subset of the given first order clauses is satisfiable, the optimisation problem would still be NP-hard. However, *every* unsatisfiable core needs to be repaired in order to make the encoding consistent, so *any* core gives potentially useful information. For that reason, it is not obvious that investing time in minimising the size of the core returned is justified. The

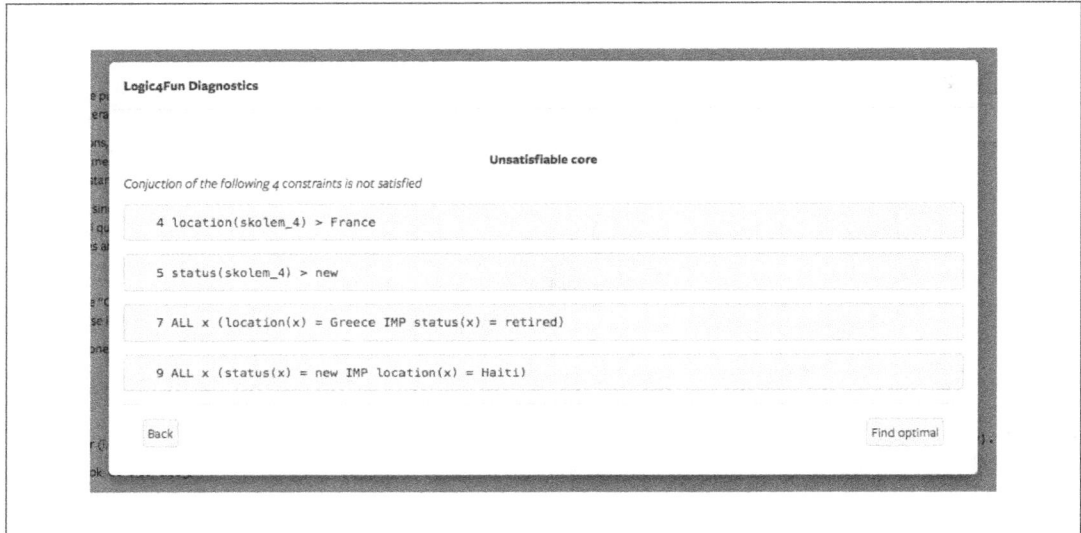

Figure 5: Screenshot of diagosis tool: unsatisfiable core

"Find optimal" button visible in Figure 5 provides an optionto seek an optimal core, but in view of the complexity issues, and because with small problem encodings the arbitrary core is quite often optimal anyway, this feature may be omitted from the tool when it is eventually deployed.

At present, the diagnoser works with the problem *after* it has been put into clause form. Consequently, the constraints violated by approximate models and those featuring in unsatisfiable cores are reported as clauses (with a little syntactic sugar such as universal quantifiers binding the variables) rather than as the input first order formulae. This is sometimes a little confusing for logically inexpert users, but is does have its advantages as it results in more precise diagnoses.

The two services provided by the diagnosis tool are in a good sense dual to each other [11]. More strictly, the set of unsatisfied clauses in an approximate model is a *diagnosis* in that it forms a hitting set for the set of all unsatisfiable cores, and dually each unsatisfiable core is *conflict*, which is a hitting set for the set of all such diagnoses. There is no general answer to the question of which is more useful, as it depends on the problem—and to some extent on the user. It may happen that there is no unsatisfiable core much smaller than the entire clause set, in which case the best strategy for the user is to ask repeatedly for approximate models, making constraints hard if they are obviously true. In other cases, the best approximate model may look nothing like the intended solution, and may violate many of the

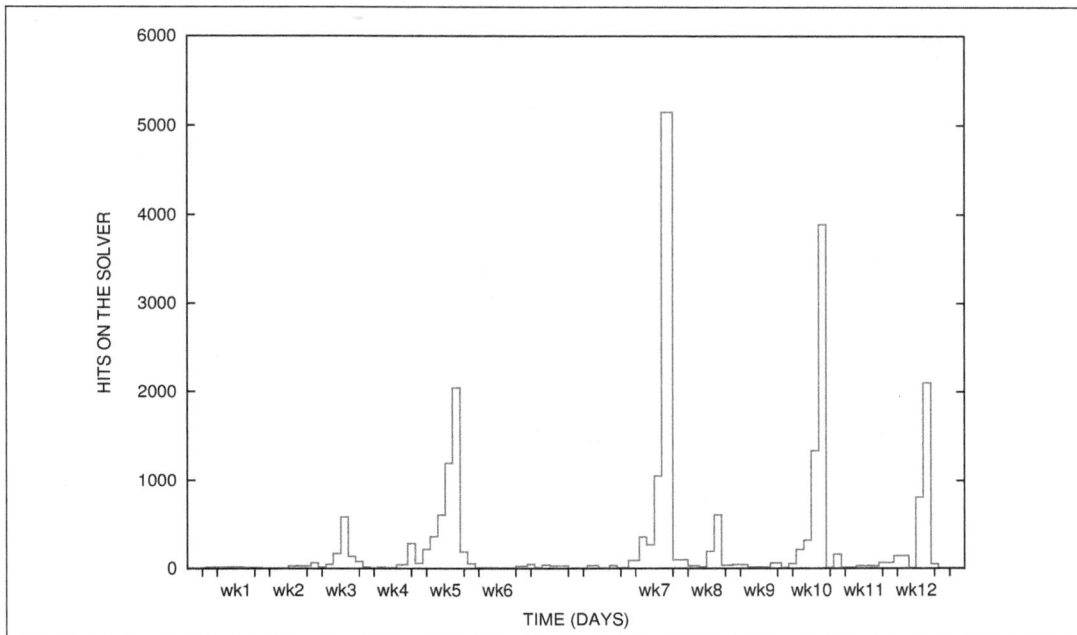

Figure 6: Number of times the solver was run on each day over 12 weeks of a logic course

constraints. In such cases it is likely that there are several bugs in the encoding, so a good technique is to isolate each one, in a small unsatisfiable core if possible, and re-examine the problem after each repair.

The diagnosis tool is not a magic bullet. Sometimes it helps; sometimes it does not. In any case, it provides only information about models or the lack of them: repairing the problem encoding is still a task for the user and still rests on understanding first order logical notation. It does, however, reduce frustration by assisting with the reasoning process rather than letting "No solution found" be a backtracking point in the dialogue between user and machine.

6 Site usage

Logfiles produced by the scripts on the site can be mined for data on usage patterns, and provide some insight into how students set about mastering the web-based tool and using it to solve problems of logic. At the simplest, aggregated statistics for the number of hits on the site allow us to observe class behaviour. Figure 6 shows the number of times the solver was run each day by a class of around 50 undergraduates

---

Logic Games

Our annual logic competition came to a final showdown between five teams: the Aces, Buccaneers, Cougars, Demons and Eagles. The contest was a round robin, each team meeting each other team once. At the end, the judges announced:

- Every team won at least once, and some team won all its games.

- The Buccaneers beat only the Cougars.

- Exactly one match was drawn, and it didn't involve the Cougars.

- The Aces defeated every team that the Eagles defeated, but they didn't defeat the Demons.

- Not every team that defeated the Aces defeated every team that the Aces defeated.

With that, they left us to work out the full set of results. Well?

Figure 7: The homework problem for week 7

during a 12-week semester in 2013. Note that there was a 2-week break between weeks 6 and 7.

Students had a piece of homework to do each week, and had to hand it in for assessment before midnight each Friday evening. These assignments in weeks 5, 7, 10 and 12 consisted of problems to be solved using *Logic for Fun*, as can clearly be seen from the bursts of activity in those weeks. The assignment in week 8 was a problem concerning the semantics of some first order formulae, which the students were asked to compute on paper, using semantic tableaux, before comparing their answers with models of the same formulae produced by *Logic for Fun*. The small peak in week 3 is associated with the point at which they were introduced to the site and asked to complete some easy exercises to familiarise themselves with it. The homework in weeks 3, 4, 6, 9 and 11 did not call for students to use the site.

The heaviest usage of the semester occurred on the Friday of week 7, when the solver was run on average almost 100 times per student. This represents an extraordinary amount of work by the class on what was a fairly modest piece of logic homework. The problem in question (see Figure 7) was designed to turn on the correct handling of quantifiers, though the most awkward part of the problem is to encode the notion of a drawn match. Since the usage log from that period only records who ran the solver at what time, not the full text of what they sent to the solver, we have no way now of knowing what particular difficulties caused the class

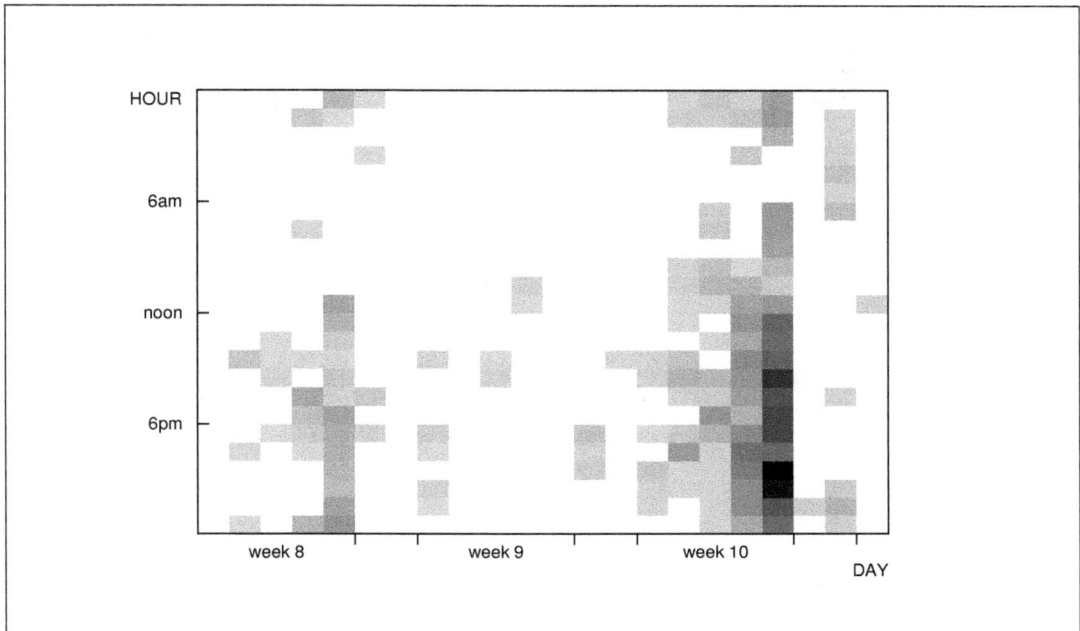

Figure 8: More detailed plot of site usage hour by hour over 3 weeks.

to spend so much more effort on this problem than on the others.

Closer analysis of the aggregate site usage reveals further patterns in students' work. Figure 8 shows the number of runs in each hour over a period of three weeks culminating in an assignment which required them to encode the problem 'Logic Games' (Figure 7). Darker colours indicate heavier usage. Although the amount of work peaks as expected on the day of the assignment deadline (not at the eleventh hour, but about three hours prior to that) there is clearly a substantial number of students whose habit is to work on problems such as these some days in advance. Recall that the "assignment" was only a piece of homework requiring a few hours of work at most, so for most students it was feasible to leave the job until the last day. The activity two days after the deadline was that of students completing the assignment late: they were allowed to submit work up to 48 hours after the deadline, for a marks penalty, and for unknown reasons some of them chose to exercise this option.

Yet more detail is revealed by studying the work patterns of individual students. Some behaviours are quite striking: there are, for example, students who will run the solver 200 times or more on one problem, many of the runs being only seconds apart. Figure 9 shows an example of the activity of one such student over the five

189

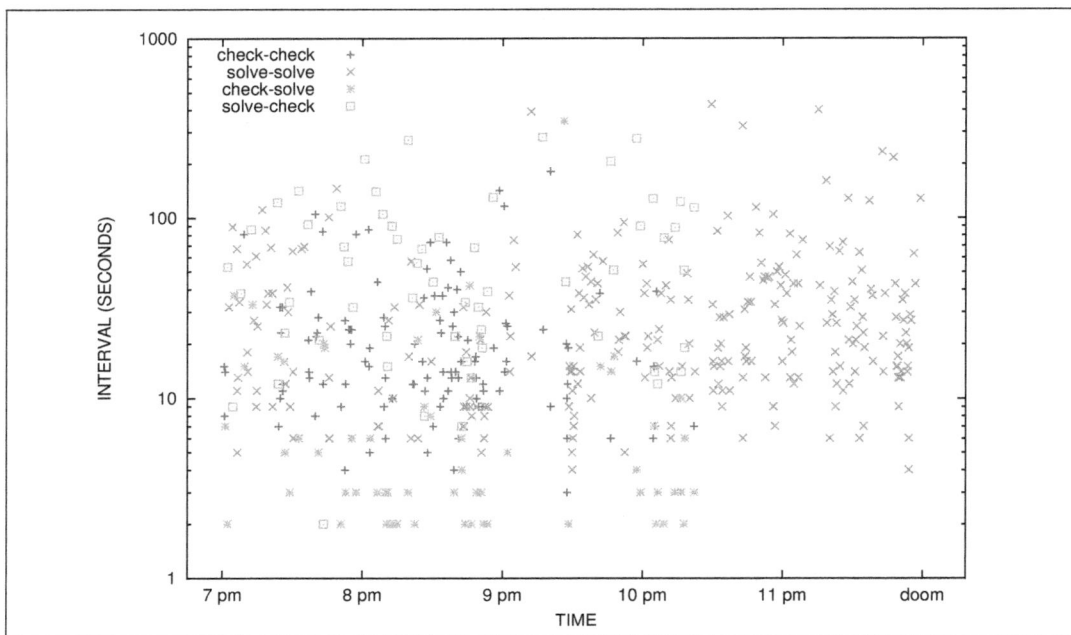

Figure 9: Five hours of work by one student. Each point is a run of the solver, showing the interval since the previous run plotted against the time of day. Different shapes show whether the run was a syntax check or a "solve", and whether the previous run was a check or a solve.

hours before the submission deadline. Note that this is a huge amount of work compared with the few minutes which students normally spend on a hand-written formalisation exercise. At no point during the five hours does this student pause for much more than 5 minutes. Most of the runs which occur within 10 seconds of the previous run are cases in which a syntax check is immediately followed by a "solve", presumably because there was no syntax error. We see some patterns in the record: for instance, at some point (a little before 10:30 pm) this student abandons explicit syntax checks and simply uses the "Solve" button. It is unclear why.

7 Current and future work

Logic for Fun was completely re-scripted in 2013–14, partly because the look and feel of the old site dated from another millennium, partly to extend its functionality in significant ways, and partly to have a version built with modern tools which would be easy to maintain. The new site is scripted entirely in Python, with a little

Javascript for client-side functions such as insertion of logical symbols into the text, though the solver behind it is still the original, written in C some 20 years ago. The beta version of the new site [10] is now freely available to all web users, in contrast to the old site which required them to have accounts and to pay fees. This is in line with the contemporary expansion of free access to educational tools.

The biggest enhancement from the user's viewpoint is a facility to save and reload work, allowing it to be carried over easily from one session to another. Students who join a class (called a "group" on the site) can also submit their work to the group manager for feedback or assessment. Low-level improvements, such as organising the display so that the natural language version of a puzzle can be in the same browser window as the student's logical encoding of it, also do much to enhance the user experience. The syntax for declaring vocabulary has been simplified and made more similar to that of analogous declarations in computational settings. Thus whereas the classic site used mathematical syntax for declarations like

```
function f: s -> t.
function P: s -> bool.
```

the new site uses "computational" syntax for the same thing:

```
function f(s): t.
predicate P(s).
```

There is an ambitious plan for continued upgrading of the site. Tools currently under development include the diagnosis assistant already noted in Section 5 above, for use when a problem as written by the user has no solution. A prototype of this tool looks promising, and has been positively received by users who have tested it, but it will not be incorporated into the live site until 2016. It is also planned that the system will maintain a detailed database of user activity, recording every character of text sent to the solver. This information will be used in a project aimed at deeper understanding of the logic learning process.

An important piece of future work, to be conducted when the new site is stable, is an effectiveness study. It is obvious that doing formalisation exercises online rather than on paper causes students to put much more effort into getting their answers right, but measuring the extent to which they learn more as a result is much harder. There is no naturally occurring control group for an experiment, so it is important that institutions other than the Australian National University begin using the site and generate comparative results between the years when it is used and years when it is not. Such longitudinal data will take time to accumulate, and no such study is available yet.

Acknowledgements   The author wishes to thank Matt Gray, Kahlil Hodgson, Daniel Pekevski, Nathan Ryan and Wayne Tsai for help in scripting *Logic for Fun*, Nursulu Kuspanova for her invaluable work on the diagnosis tool, and the many logic students over the last 15 years who have helped by testing the site to destruction.

References

[1] Dave Barker-Plummer, Jon Barwise, and John Etchemendy. Tarski's world, 2008.

[2] Jon Barwise and John Etchemendy. Tarski's world, 1993.

[3] Jon Barwise and John Etchemendy. *Language, Proof and Logic.* CSLI Publications, Stanford, CA, USA, 1999.

[4] Maria Garcia de la Banda, Kim Marriott, Reza Rafeh, and Mark Wallace. The modelling language Zinc. In *Principles and Practice of Constraint Programming (CP)*, pages 700–705, 2006.

[5] Wilfred Hodges. *Logic: An introduction to elementary logic, 2nd edn.* Penguin, London, 2005.

[6] Kim Marriott, Nicholas Nethercote, Reza Rafeh, Peter Stuckey, Maria Garcia de la Banda, and Mark Wallace. The design of the Zinc modelling language. *Constraints*, 13:229–267, 2008.

[7] Nicholas Nethercote, Peter Stuckey, Ralph Becket, Sebastian Brand, Gregory Duck, and Guido Tack. MiniZinc: Towards a standard CP modelling language. In *Principles and Practice of Constraint Programming (CP)*, pages 529–543, 2007.

[8] Bart Selman, Henry Kautz, and Bram Cohen. Local search strategies for satisfiability testing. *Theoretical Computer Science*, DIMACS Series Volume: Clique, Graph coloring and Satisfiability—Second DIMACS implementation challenge. American Math Soc:290–295, 1996.

[9] John Slaney. Logic for Fun (classic version). `http://logic4fun.rsise.anu.edu.au/`.

[10] John Slaney. Logic for Fun, Version 2 Beta. `http://{L4F}.cecs.anu.edu.au/`.

[11] John Slaney. Set-theoretic duality: A fundamental feature of combinatorial optimisation. In *European Conference on Artificial Intelligence (ECAI)*, pages 843–848, 2014.

[12] Peter Stuckey and Carleton Coffrin. Modelling discrete optimization. `https://www.class-central.com/mooc/3692/coursera-modeling-discrete-optimization`.

                                           Received 11 October 2016

Teaching natural deduction in the right order with Natural Deduction Planner

Declan Thompson
The University of Auckland, New Zealand
declanthompsonnz@gmail.com

Jeremy Seligman
The University of Auckland, New Zealand
j.seligman@auckland.ac.nz

Abstract

We describe a strategy-based approach to teaching natural deduction using a notation that emphasises the order in which deductions are constructed, together with a LaTeX package and Java app to aid in the production of teaching resources and classroom demonstrations. Our approach is aimed at students with little exposure to mathematical method and has been developed while teaching undergraduate classes for philosophy students over the last ten years.

Keywords: Natural Deduction, Strategy, Proof Assistant.

1 Natural Deduction as a Creative Process

Teaching modern logic to students with little background in mathematics is notoriously hard. The philosophy student, adept at reading complex prose and composing artful essays is usually not well prepared for manipulating symbols and constructing rigorous proofs of theorems. Acquisition of at least the following three skills are needed.

The first is using the language of propositional and predicate logic to represent one's thoughts in formal notation and understand what has been written by others. This is usually achieved by learning to translate to and from natural language. Many resources are available.

The second is manipulating the symbols of formal notation according to precise rules. This is a basic skill necessary for almost all of logic, from applying mechanical methods of argument evaluation to acquiring an appreciation of the autonomy of the syntactic realm, without which the major theoretical results of logical theory cannot be understood. It can be acquired by learning how to produce truth tables, truth trees, and in many other ways. Again many resources are available.

The third is reading and writing rigorous arguments, of the kind used in mathematics. This is a much more difficult skill to acquire, requiring mastery of the first two skills, and in addition, a level of mathematical maturity that is attained by mathematics students only after years of practice with algebra, geometry, analysis, etc. Consequently, this side of logic education is often neglected by philosophy undergraduate programmes. Although many introductory textbooks include some discussion of logical theory, such as soundness and completeness, the emphasis is on understanding the theorems rather than developing the skills to prove them. Few are aimed directly at acquiring the skill of creating proofs from scratch.[1]

One solution is to require logic students to take a substantial number of courses in mathematics, so that they acquire the necessary skills in the same way as mathematics students. In the long run, a broad experience with mathematical methods is certainly useful for research in logic, if not absolutely essential. But the huge gap that must be filled is daunting and dispiriting for most philosophy students, most of whom decide that it is just too big to breach.

Is there another solution? The obvious candidate is to teach students the skill of rigorous argumentation using the very formalisms that they have already learned: propositional and predicate logic. From a theoretical perspective, we know that our various systems of deduction can duplicate all that a mathematics student learns by a much more indirect and less explicit route. Why then is it so difficult for a philosophy student who has learned a formal system of deduction to transfer her skills to the production of informally rigorous arguments of the kind needed for progress in her subject?

It is generally recognised that axiomatic systems, while elegant and theoretically parsimonious, are wholly inappropriate for learning deduction. Instead, most logic programmes for philosophy students include some system of natural deduction, in which axioms are replaced by rules which mirror patterns of reasoning used in natural language argumentation. In the classic approach of Irving Copi, numerous rules are added, so as to capture as many such patterns as possible.[2] Yet there is often an insufficient level of attention to any systematic discussion of the process of *creating*

---

[1]A notable exception is "How to Prove It: A Structured Approach" [1, 2]. There are also countless introductions to mathematical method, e.g. "The Nuts and Bolts of Mathematical Proof" [3], "How to Read and Do Proofs: An Introduction to Mathematical Thought Processes" [4].

[2]"Introduction to Logic" [5] is now in its 14th edition. Many other textbooks on natural deduction employ a similarly lengthy list of rules.

deductions. Typically, students are given an introduction to the rules, motivated by their natural language correlates, a few examples of complete deductions, and are then left to fend for themselves on a large number of exercises, with the hope that they will develop their own strategies by trial and error.

An alternative is to teach the *strategies* of creation explicitly. As well as helping students to learn formal deduction, these are the strategies that will prepare the student for the harder task of creating rigorous informal arguments of the kind needed to do postgraduate work in logic, and which mathematics students learn implicitly through their application to a wide range of mathematical topics. Teachers of natural deduction in the traditional style may be fully aware of this point, but the effective learning of explicit strategy is made almost impossibly hard by several factors.

The first is simply the number of rules used by logic textbooks aimed at mirroring patterns in informal reasoning, which include both proof by cases (Disjunction Elimination, $\vee\mathbf{E}$), and Disjunctive Syllogism, if not also Constructive Dilemma. While each of these is relatively easy to explain in isolation, the more rules, the harder it is to master their strategic interactions, which the student must consider when creating her own deductions.

A second, related factor is the lack of structure to the set of rules. From the perspective of teaching strategy, one would prefer a simpler set of rules, organised in a way that corresponds to patterns of use in the creation of deductions, and exactly this is provided by Gentzen's original system, which uses the idea of introduction and elimination to expose the structure and symmetry of proof. More details will be given in Section 2, but for now a brief summary of the main points will suffice. Firstly, the fact that the intuitionistic fragment of the system has only a pair of rules for each logical operator allows one to develop general strategies: one for Introduction rules and one for Elimination rules, concerning the management of resources and simplification of goals. Moreover, an orthogonal classification of rules allows us to distinguish between cases in which a choice is required (e.g., $\vee\mathbf{I}$ and $\exists\mathbf{I}$) and those that are 'automatic', in the sense that they can be applied without the need for further choice. Even the symmetry-breaking oddity of the non-intuitionistic rule $\neg\neg\mathbf{E}$, which can be applied to any conclusion, raises an important strategic question: how to manage the creative steps of deduction? And this leads to an explicit discussion of back-tracking in problem solving and the need to recognise dead-ends. While such matters of strategy are implicit in more complicated systems, they are highlighted in systems in which the number of rules is small and well-balanced.

Standard presentations of Gentzen-style natural deduction, such as those of Fitch or Lemon, still have an important deficiency. Designed for reading rather than writing, the argument is displayed with the premises at the top, the conclusion at the bottom and with each line justified by lines higher up on the page, according to a formal rule. This makes the process of checking the correctness of the deduction relatively easy, but the process of generating the deduction itself unnecessarily hard.

1.	$(\neg p \vee \neg q)$	**Prem**	*On the left is a correct deduction*
2.	$(p \wedge q)$	**Ass**	*using a version of Gentzen's rules.*
3.	$\neg p$	**Ass**	*Hypothetical reasoning is indicated*
4.	p	$2, \wedge$**E**	*by marking the assumption (**Ass**)*
5.	$\bot$	$3, 4, \neg$**E**	*and a vertical bracket ending be-*
6.	$\neg q$	**Ass**	*low the hypothetical conclusion. The*
7.	q	$2, \wedge$**E**	*symbol $\bot$ is used to mark a contra-*
8.	$\bot$	$6, 7, \neg$**E**	*diction.*
9.	$\bot$	$1, 3\text{-}5, 6\text{-}8, \vee$**E**	
10.	$\neg(p \wedge q)$	$2\text{-}9, \neg$**I**	

Information about the process of *creating* the deduction is lost in this representation, which wrongly suggests that it was written from top to bottom, starting with the premises and ending in the conclusion. (One of the most common mistakes made by students is to follow this order.) There is no record of the strategies used to construct the deduction; no record even of the order in which it was constructed. The student who fails to produce her own deduction of $\neg(p \wedge q)$ from $(\neg p \vee \neg q)$ will not learn much from looking at the above solution.

If we were to display a full sequence of steps leading to the creation of this deduction, we might write the following:

This is much too cumbersome for practical use in textbooks, and leaves the assignment of line numbers somewhat mysterious. How to know the deduction will use ten lines? But a simple change in notation can help. Instead of numbering the lines of the deduction from top to bottom, we number them in the order they were created. The above sequence can then be represented with just one deduction, as shown:

1.	$(\neg p \vee \neg q)$	**Prem**	
3.	$(p \wedge q)$	**Ass**	
5.	$\neg p$	**Ass**	
9.	p	$3, \wedge \mathbf{E}$	
6.	$\bot$	$5, 9, \neg \mathbf{E}$	
7.	$\neg q$	**Ass**	
10.	q	$3, \wedge \mathbf{E}$	
8.	$\bot$	$7, 10, \neg \mathbf{E}$	
4.	$\bot$	$1, 5\text{-}6, 7\text{-}8, \vee \mathbf{E}$	
2.	$\neg (p \wedge q)$	$3\text{-}4, \neg \mathbf{I}$	

First, the premise and conclusion are written as lines 1 and 2, with a generous space between. We then apply $\neg \mathbf{I}$ to the conclusion to get a hypothetical deduction with assumption $(p \wedge q)$ on line 3 and conclusion $\bot$ on line 4. Next, we apply $\vee \mathbf{E}$ to line 1 to get two nested hypothetical deductions, from $\neg p$ on line 5 to $\bot$ on line 6, and from $\neg q$ on line 7 to $\bot$ on line 8. The first of these is completed using $\neg \mathbf{I}$ to get p on line 9 (justified by $\wedge \mathbf{E}$ from line 3). The second is completed similarly, with line q on line 10. In this way, the line numbers match the order of construction of the deduction precisely, which is thereby emphasised to students as they create it.

The discipline of numbering in the order a deduction is created helps students (and instructors) to think strategically. The goal is to provide a justification for the conclusion given the resources in the premises, and seen this way deduction is just planning how to use the resources to satisfy a goal. While this is a familiar idea in automated reasoning research, it rarely enters the classroom. By using the above system of numbering, students cannot avoid thinking in this strategic way and learning that introduction rules serve to split the goal into subgoals, whereas elimination rules deploy resources. Strategic concepts such as back-tracking, management of decision points, and an awareness of risk are brought to the fore. Certain rules, such as Disjunction Introduction are seen as "choice rules" to be used with caution and postponed as long as possible, whereas others, such as Implication Introduction are "automatic" - they can and should be used immediately with no risk of having to undo.

The use of a new notation has the disadvantage that teaching resources, especially solutions to exercises, have to be produced from scratch. And it was to aid in this that we decided to produce both a LaTeX package for formatting our deductions easily, and a Java app to aid in the generation of both LaTeX code and various other formats for classroom demonstration.

2 Teaching Strategies, Explicitly

At any stage of creating a deduction, one has to decide which rule to apply next. One of the great advantages of using Gentzen-style systems of natural deduction is both the relative paucity of rules and the direction one has from the syntactic structure of premises and conclusion as to which rule to use. For intuitionistic natural deduction, whose rules in Fitch-style are shown in Table 1, exactly one rule applies to each formula and this fact makes it easy to teach specific strategies.

2.1 Goals and Resources

The first distinction we teach is that between "goals" and "resources". When a natural deduction problem is first written down in Fitch-style, it consists of a number of lines at the top (the premises) and a line at the bottom (the conclusion). For example:

$$
\begin{array}{lll}
1. & (p \wedge q) & \textbf{Prem} \\
2. & (p \rightarrow r) & \textbf{Prem} \\
  & \vdots & \\
3. & (p \rightarrow (q \wedge r)) &
\end{array}
$$

Here, the *goal* is to prove line 3, $(p \rightarrow (q \wedge r))$ from *resources* on lines 1 and 2. We can write this explicitly in sequent notation as

$$(p \wedge q), (p \rightarrow r) \implies (p \rightarrow (q \wedge r))$$

Students are not taught sequent notation initially. Instead, we discuss the role of goals and resources by referring directly to Fitch-style deductions, written on the whiteboard or projected from a computer. Nonetheless, they are useful here as a way of making proof strategies explicit.

As a deduction progresses, one typically has multiple goals and resources. For example, in the deduction shown below, we have two remaining tasks:

Natural Deduction, NJ			
Conjunction Introduction $i \quad \phi$ $j \quad \psi$ $\Rightarrow \quad (\phi \wedge \psi) \quad i,j,\wedge\mathbf{I}$	**Conjunction Elimination** $i \quad (\phi \wedge \psi)$ $\Rightarrow \quad \phi \quad i,\wedge\mathbf{E}$ .... $i \quad (\phi \wedge \psi)$ $\Rightarrow \quad \psi \quad i,\wedge\mathbf{E}$	**Implication Introduction** $\ulcorner \; i \quad \phi \qquad$ **Ass** $\quad\;\; \vdots$ $\llcorner \; j \quad \psi$ $\Rightarrow \quad (\phi \rightarrow \psi) \quad i{-}j,\rightarrow\mathbf{I}$	**Implication Elimination** $i \quad (\phi \rightarrow \psi)$ $j \quad \phi$ $\Rightarrow \quad \psi \quad i,j,\rightarrow\mathbf{E}$
Disjunction Introduction $i \quad \phi$ $\Rightarrow \quad (\phi \vee \psi) \quad i,\vee\mathbf{I}$ .... $i \quad \psi$ $\Rightarrow \quad (\phi \vee \psi) \quad i,\vee\mathbf{I}$	**Disjunction Elimination** $i \quad (\phi \vee \psi)$ $\ulcorner \; j \quad \phi \qquad$ **Ass** $\quad\;\; \vdots$ $\llcorner \; k \quad \theta$ $\ulcorner \; l \quad \psi \qquad$ **Ass** $\quad\;\; \vdots$ $\llcorner \; m \quad \theta$ $\Rightarrow \quad \theta \quad i,j{-}k,l{-}m,\vee\mathbf{E}$	**Equivalence Introduction** $\ulcorner \; i \quad \phi \qquad$ **Ass** $\quad\;\; \vdots$ $\llcorner \; j \quad \psi$ $\ulcorner \; k \quad \psi \qquad$ **Ass** $\quad\;\; \vdots$ $\llcorner \; l \quad \phi$ $\Rightarrow \quad (\phi \leftrightarrow \psi) \quad i{-}j,k{-}l,\leftrightarrow\mathbf{I}$	**Equivalence Elimination** $i \quad (\phi \leftrightarrow \psi)$ $j \quad \phi$ $\Rightarrow \quad \psi \quad i,j,\leftrightarrow\mathbf{E}$ .... $i \quad (\phi \leftrightarrow \psi)$ $j \quad \psi$ $\Rightarrow \quad \phi \quad i,j,\leftrightarrow\mathbf{E}$
Negation Introduction $\ulcorner \; i \quad \phi \qquad$ **Ass** $\quad\;\; \vdots$ $\llcorner \; j \quad \bot$ $\Rightarrow \quad \neg\phi \quad i{-}j,\neg\mathbf{I}$	**Negation Elimination** $i \quad \neg\phi$ $j \quad \phi$ $\Rightarrow \quad \bot \quad i,j\neg\mathbf{E}$	**Falsum Introduction**	**Falsum Elimination** $i \quad \bot$ $\Rightarrow \quad \phi \quad i,\bot\mathbf{E}$
Universal Introduction $i \quad \phi^x_a$ $\Rightarrow \quad \forall x\,\phi \quad i,\forall\mathbf{I}$ a is a parameter not in ϕ, nor any premises, nor active assumptions	**Universal Elimination** $i \quad \forall x\,\phi$ $\Rightarrow \quad \phi^x_t \quad i,\forall\mathbf{E}$ t is any term	**Existential Introduction** $i \quad \phi^x_t$ $\Rightarrow \quad \exists x\,\phi \quad i,\exists\mathbf{I}$ t is any term	**Existential Elimination** $i \quad \exists x\,\phi$ $\ulcorner \; j \quad \phi^x_a \qquad$ **Ass** $\quad\;\; \vdots$ $\llcorner \; k \quad \psi$ $\Rightarrow \quad \psi \quad i,j{-}k\exists\mathbf{E}$ a is a parameter not in ϕ, ψ, nor any premises, nor active assumptions

Table 1: The rules of intuitionistic natural deduction (NJ) in a standard presentation.

199

$$
\begin{array}{lll}
1. & ((p \wedge q) \to r) & \textbf{Prem} \\
2. & (q \vee \neg(p \wedge r)) & \textbf{Prem} \\
4. & p & \textbf{Ass} \\
6. & q & \textbf{Ass} \\
& \quad \vdots & \\
7. & r & \\
8. & r & \textbf{Ass} \\
10. & q & \textbf{Ass} \\
11. & q & 10 \\
12. & \neg(p \wedge r) & \textbf{Ass} \\
& \quad \vdots & \\
13. & q & \\
9. & q & 2,\ 10\text{-}11,\ 12\text{-}13,\ \vee\textbf{E} \\
5. & (q \leftrightarrow r) & 6\text{-}7, 8\text{-}9,\ \leftrightarrow \textbf{I} \\
3. & (p \to (q \leftrightarrow r)) & 4\text{-}5,\ \to \textbf{I}
\end{array}
$$

These can be represented explicitly by listing the resources and goal in sequent form:

From 1,2,4,6 to 7: $((p \wedge q) \to r), (q \vee \neg(p \wedge r)), p, q \implies r$

From 1,(2),4,8,12 to 13: $((p \wedge q) \to r), p, r, \neg(p \wedge r) \implies q$

Line 2 has been dropped as a resource for the second task because it has already been used in the $\vee \textbf{E}$ used to justify line 9. Learning to identify the remaining goals of a deduction is relatively easy: they are simply the line which still lack a justification. Learning the available resources requires an understanding of the nesting of hypothesis lines, but is more apparent in the process of constructing the deduction than in reading it.

2.2 Automatic Rules

Our second main distinction is between "automatic" application of rules and applications that involve a "choice". A paradigmatic example of an automatic rule application is any use of $\to \textbf{I}$. Here, the goal is an implication and the deduction is transformed by adding its antecedent as a resource and making its succedent the new goal:

$$
\begin{array}{lll}
1. & (p \wedge q) & \textbf{Prem} \\
2. & (p \to r) & \textbf{Prem} \\
4. & p & \textbf{Ass} \\
& \quad \vdots & \\
5. & (q \wedge r) & \\
3. & (p \to (q \wedge r)) & 4\text{-}5,\ \to \textbf{I}
\end{array}
$$

We can represent an application of $\to \textbf{I}$ in sequent form as:

$\rightarrow \mathbf{I}$	$\leftrightarrow \mathbf{I}$	$\neg \mathbf{I}$
$$\frac{\Gamma, \phi \implies \psi}{\Gamma \implies (\phi \rightarrow \psi)}$$	$$\frac{\Gamma, \phi \implies \psi \qquad \Gamma, \psi \implies \phi}{\Gamma \implies (\phi \leftrightarrow \psi)}$$	$$\frac{\Gamma, \phi \implies \bot}{\Gamma \implies \neg A}$$
$\exists \mathbf{E} \quad (a \text{ new})$	$\vee \mathbf{E}$	$\forall \mathbf{I} \quad (a \text{ new})$
$$\frac{\Gamma, \phi_a^x \implies \psi}{\Gamma, \exists x\, A \implies \psi}$$	$$\frac{\Gamma, \phi \implies \theta \qquad \Gamma, \psi \implies \theta}{\Gamma, (\phi \vee \psi) \implies \theta}$$	$$\frac{\Gamma \implies \phi_a^x}{\Gamma \implies \forall x\, \phi}$$
$\bot \mathbf{E}$	$\wedge \mathbf{I}$	$\wedge \mathbf{E}$
$$\frac{}{\Gamma, \bot \implies \phi}$$	$$\frac{\Gamma \implies \phi \qquad \Gamma \implies \psi}{\Gamma \implies (\phi \wedge \psi)}$$	$$\frac{\Gamma, \phi, \psi \implies \theta}{\Gamma, (\phi \wedge \psi) \implies \theta}$$

Table 2: The automatic strategy rules of **NJ**

$$\frac{\Gamma, \phi \implies \psi}{\Gamma \implies (\phi \rightarrow \psi)}$$

Reading upwards, this means that the deduction task with goal $(\phi \rightarrow \psi)$ and resources Γ is replaced by the task with goal ψ and resources Γ, ϕ. The rule application is *automatic* because of the equivalence of these two tasks: ψ is a consequence of Γ, ϕ if and only if $(\phi \rightarrow \psi)$ is a consequence of Γ. The application of the rule usually also results in a decrease in the complexity of the deduction, splitting up the formula $(\phi \rightarrow \psi)$ into its proper parts. Students are therefore encouraged to apply the rule of $\rightarrow \mathbf{I}$ automatically, without the need to think strategically.

All the hypothetical rules are similarly automatic: $\rightarrow \mathbf{I}$, $\leftrightarrow \mathbf{I}$, $\neg \mathbf{I}$, $\exists \mathbf{E}$ and $\vee \mathbf{E}$. In addition, among the non-hypothetical rules, $\forall \mathbf{I}$, $\bot E$, $\wedge \mathbf{I}$ and $\wedge \mathbf{E}$ are automatic. Sequent notation for these rules in shown in Table 2. Each of the automatic rules satisfies a similar equivalence: the sequent below the line is valid iff those above the line are all valid. Strategically, this means that nothing is lost by applying the rule. If the task below the line is completable, then it can be completed by applying the rule and moving to the task(s) above the line. There is no risk of failure.

The majority of rules in **NJ** are automatic and so students have a lot of guidance as to how to proceed. After a certain amount of practice, they can easily be trained to identify the automatic rules and apply them almost mechanically, giving them confidence in following the goal-resource methodology rather than simply forward chaining from the premises in an attempt to reach the conclusion. This breaking of bad habits and providing a reliable framework within which more complex strategies can be developed is a large part of what we can do by teaching natural deduction "in the right order".

2.3 Choice and Procrastination

The automatic application of rules is contrasted with the application of rules that require a choice. A paradigmatic example of a choice rule is $\vee\mathbf{I}$. Faced with the goal of proving $(\phi \vee \psi)$, we have to decide which: ϕ or ψ. For example, in the problem on the left, an application of $\vee\mathbf{I}$ with choice p quickly leads to impasse (right):

1.	$((p \to r) \leftrightarrow (r \to q))$	**Prem**		1.	$((p \to r) \leftrightarrow (r \to q))$	**Prem**
2.	r	**Prem**		2.	r	**Prem**
				7.	r	**Ass**
				8.	q	
				5.	$(r \to q)$	7-8, $\to \mathbf{I}$
				6.	$(p \to r)$	1, 5, $\leftrightarrow \mathbf{E}$
				4.	p	
3.	$(p \vee q)$			3.	$(p \vee q)$	4, $\vee\mathbf{I}$

Students are presented with examples and exercises like this one to show that a real choice is made and that one must recognise when one is stuck so as to backtrack to the last choice. In this case, one can simply make the other choice, q and complete the deduction:

1.	$((p \to r) \leftrightarrow (r \to q))$	**Prem**
2.	r	**Prem**
7.	p	**Ass**
8.	r	2
5.	$(p \to r)$	7-8, $\to \mathbf{I}$
6.	$(r \to q)$	1, 5, $\leftrightarrow \mathbf{E}$
4.	q	6, 2, $\to \mathbf{E}$
3.	$(p \vee q)$	4, $\vee\mathbf{I}$

In sequent notation, the choice is displayed vividly as a choice between two rules:

$$\frac{\Gamma \implies \phi}{\Gamma \implies (\phi \vee \psi)} \qquad \frac{\Gamma \implies \psi}{\Gamma \implies (\phi \vee \psi)}$$

Neither shares the equivalence of the automatic rules. It is possible for one or more of sequents above the line to be invalid (and so not provable) even if the sequent below the line is provable, by some other means. So in fact, the situation is worse than merely choosing between ϕ or ψ when trying to prove $(\phi \vee \psi)$. It may be that neither choice works:

202

1.	$(r \lor s)$	**Prem**
2.	$(r \to p)$	**Prem**
3.	$\neg(s \land q)$	**Prem**
4.	$(p \lor \neg q)$	

Here, applying $\lor$**I** at step 5 is no good no matter which disjunct we choose:

1.	$(r \lor s)$	**Prem**
2.	$(r \to p)$	**Prem**
3.	$\neg(s \land q)$	**Prem**
6.	r	**Ass**
7.	p	2, 6, $\to$ **E**
8.	s	**Ass**
	$\vdots$	
12.	q	
10.	$(s \land q)$	8, 12, $\land$**I**
11.	$\perp$	3, 10, $\neg$**E**
9.	p	11, $\perp$**E**
5.	p	1, 6-7, 8-9, $\lor$**E**
4.	$(p \lor \neg q)$	5, $\lor$**I**

1.	$(r \lor s)$	**Prem**
2.	$(r \to p)$	**Prem**
3.	$\neg(s \land q)$	**Prem**
6.	q	**Ass**
9.	r	**Ass**
14.	p	2, 9, $\to$ **E**
	$\vdots$	
13.	s	
10.	$(s \land q)$	13, 6, $\land$**I**
11.	s	**Ass**
12.	$(s \land q)$	11, 6, $\land$**I**
8.	$(s \land q)$	1, 9-10, 11-12, $\lor$**E**
7.	$\perp$	3, 8, $\neg$**E**
5.	$\neg q$	6-7, $\neg$**I**
4.	$(p \lor \neg q)$	5, $\lor$**I**

The only solution is to postpone the choice (leaving the $\lor$**I** to steps 9 and 10):

1.	$(r \lor s)$	**Prem**
2.	$(r \to p)$	**Prem**
3.	$\neg(s \land q)$	**Prem**
5.	r	**Ass**
9.	p	2, 5, $\to$ **E**
6.	$(p \lor \neg q)$	9, $\lor$**I**
7.	s	**Ass**
11.	q	**Ass**
13.	$(s \land q)$	7, 11, $\land$**I**
12.	$\perp$	3, 13, $\neg$**E**
10.	$\neg q$	11-12, $\neg$**I**
8.	$(p \lor \neg q)$	10, $\lor$**I**
4.	$(p \lor \neg q)$	1, 5-6, 7-8, $\lor$**E**

Teaching how to manage one's choices and to realise that the best strategy is often to procrastinate is a core part of our approach to natural deduction. We strongly emphasise the distinction between automatic and choice rules, so as to highlight when choices are made and the nature of the choice. The pure choice rules are $\lor$**I** and $\exists$**I**, as shown in Table 3. The latter requires a choice of term t with which to replace the bound variable x of $\exists x\,\phi$. This should be taken from terms already occurring in the deduction, or if there is none, a new individual constant. In the

$\vee \mathbf{I_l}$	$\vee \mathbf{I_r}$	$\exists \mathbf{I}$ (t old/first)
$\dfrac{\Gamma \implies \phi}{\Gamma \implies (\phi \vee \psi)}$	$\dfrac{\Gamma \implies \phi}{\Gamma \implies (\phi \vee \psi)}$	$\dfrac{\Gamma \implies \phi_t^x}{\Gamma \implies \exists x\, \phi}$

Table 3: Pure choice rules of **NJ**

latter case, since any new constant will do, students may be tricked into thinking that no choice is involved, but that's not so. Just as in $\vee\mathbf{I}$, there are examples in which only the avoidance of any choice (procrastination) will enable a solution:

1.	$\exists x\,(Fx \wedge Gx)$	**Prem**	1.	$\exists x\,(Fx \wedge Gx)$	**Prem**	1.	$\exists x\,(Fx \wedge Gx)$	**Prem**
	$\vdots$		4.	$(Fb \wedge Gb)$	**Ass**	3.	$(Fa \wedge Ga)$	**Ass**
			6.	Fb	4, $\wedge\mathbf{E}$	5.	Fa	3, $\wedge\mathbf{E}$
			7.	Gb	4, $\wedge\mathbf{E}$	4.	$\exists x\, Fx$	5, $\exists\mathbf{I}$
				$\vdots$		2.	$\exists x\, Fx$	1, 3-4, $\exists\mathbf{E}$
			5.	Fa				
			3.	Fa	1, 4-5, $\exists\mathbf{E}$			
2.	$\exists x\, Fx$		2.	$\exists x\, Fx$	3, $\exists\mathbf{I}$			

The initial problem, shown on the left, is not solved by $\exists\mathbf{I}$, shown in the middle, despite there being no choice of instantiating term; only by procrastination can the deducting be completed (right).

2.4 Rules of Deduction vs Rules of Strategy

While the distinction between automatic application of rules and those that require choice management lines up with the natural deduction rules in all the cases considered above, the three remaining rules, of $\to\mathbf{E}$, $\leftrightarrow\mathbf{E}$ and $\neg\mathbf{E}$ have varying strategic properties depending on the context. Applications of $\to\mathbf{E}$, for example, can be split into three cases, indicated below:

1.	$(\phi \to \psi)$		1.	$(\phi \to \psi)$		1.	$(\phi \to \psi)$	
2.	ϕ			$\vdots$			$\vdots$	
	$\vdots$		2.	ψ		2.	θ	
3.	θ							

						1.	$(\phi \to \psi)$	
1.	$(\phi \to \psi)$		1.	$(\phi \to \psi)$			$\vdots$	
2.	ϕ			$\vdots$		3.	ϕ	
4.	ψ	1, 2, $\to\mathbf{E}$				4.	ψ	1, 3, $\to\mathbf{E}$
	$\vdots$		3.	ϕ			$\vdots$	
3.	θ		2.	ψ	1, 3, $\to\mathbf{E}$	2.	θ	

Top left is an instance of *modus ponens*, in which both $(\phi \to \psi)$ and ϕ are available resources and the inference to ψ (shown below, bottom left) is automatic. Top middle is an instance of *reverse modus ponens*, in which $(\phi \to \psi)$ is a resource and the goal is ψ, and the replacement of this goal by ϕ (shown below, bottom middle) is a matter of choice; ϕ may not be provable even if ψ is by another route. Finally, when neither ϕ is a resource nor ψ a goal (top right), the only way of using the resource $(\phi \to \psi)$ is to take a *bold step* (bottom right), which is again a choice that may lead to impasse.

The introduction of this terminology ("modus ponens", "reverse modus ponens" and "bold step") helps students distinguish between rules of strategy and the actual rules of logic as defined in the system of deduction. A complete list, in sequent notation is given in Table 4. To repeat: the sequent style is not used in the classroom at this level. We teach all of these strategies through explicit examples. NDP aids significantly in this process, since the mechanics of applying the rules is performed by the software, allowing the student (or the instructor using the software) to focus on strategic matters. For example, the strategic rule of $\perp$ *bold step* is illustrated by an example like this:

1.	$(p \vee q)$	**Prem**
2.	$\neg(p \wedge r)$	**Prem**
3.	$(q \to s)$	**Prem**
5.	r	**Ass**
7.	p	**Ass**
	$\vdots$	
8.	s	
9.	q	**Ass**
10.	s	$3, 9, \to$ **E**
6.	s	$1, 7\text{-}8, 9\text{-}10, \vee$**E**
4.	$(r \to s)$	$5\text{-}6, \to$ **I**

The remaining task is $\neg(p \wedge r)$, $(q \to s)$, r, $p \implies s$. We have two complex resources available: $\neg(p \wedge r)$ and $(q \to s)$. Using $(q \to s)$ would involve reverse modus ponens, which turns out not to work. The only hope is to use $\neg(p \wedge r)$ which requires $\neg$**E**. But we have neither $(p \wedge r)$ as a resource (EFQ) nor $\perp$ as a goal ($\perp$ intro) so we have to perform a $\perp$ bold step:

Automatic	Choice	
$\to \mathbf{E}$ (MP)	$\to \mathbf{E}$ (reverse MP)	$\to \mathbf{E}$ (bold step)
$$\dfrac{\Gamma, \phi, \psi \implies \theta}{\Gamma, (\phi \to \psi), \phi \implies \theta}$$	$$\dfrac{\Gamma, (\phi \to \psi) \implies \phi}{\Gamma, (\phi \to \psi) \implies \psi}$$	$$\dfrac{\Gamma, (\phi \to \psi) \implies \phi \quad \Gamma, \psi \implies \theta}{\Gamma, (\phi \to \psi) \implies \theta}$$
$\leftrightarrow \mathbf{E}$ (L-R MP)	$\leftrightarrow \mathbf{E}$ (L-R reverse MP)	$\leftrightarrow \mathbf{E}$ (L-R bold step)
$$\dfrac{\Gamma, \phi, \psi \implies \theta}{\Gamma, (\phi \leftrightarrow \psi), \phi \implies \theta}$$	$$\dfrac{\Gamma, (\phi \leftrightarrow \psi) \implies \phi}{\Gamma, (\phi \leftrightarrow \psi) \implies \psi}$$	$$\dfrac{\Gamma, (\phi \leftrightarrow \psi) \implies \phi \quad \Gamma, \psi \implies \theta}{\Gamma, (\phi \leftrightarrow \psi) \implies \theta}$$
$\leftrightarrow \mathbf{E}$ (R-L MP)	$\leftrightarrow \mathbf{E}$ (R-L reverse MP)	$\leftrightarrow \mathbf{E}$ (R-L bold step)
$$\dfrac{\Gamma, \psi, \phi \implies \theta}{\Gamma, (\phi \leftrightarrow \psi), \psi \implies \theta}$$	$$\dfrac{\Gamma, (\phi \leftrightarrow \psi) \implies \psi}{\Gamma, (\phi \leftrightarrow \psi) \implies \phi}$$	$$\dfrac{\Gamma, (\phi \leftrightarrow \psi) \implies \psi \quad \Gamma, \phi \implies \theta}{\Gamma, (\phi \leftrightarrow \psi) \implies \theta}$$
$\neg \mathbf{E}$ (EFQ)	$\to \mathbf{E}$ ($\bot$ intro)	$\to \mathbf{E}$ ($\bot$ bold step)
$$\dfrac{\overline{\Gamma, \bot \implies \theta}}{\Gamma, \neg\phi, \phi \implies \theta}$$	$$\dfrac{\Gamma \implies \phi}{\Gamma, \neg\phi \implies \bot}$$	$$\dfrac{\Gamma \implies \phi \quad \overline{\Gamma, \bot \implies \theta}}{\Gamma, \neg\phi \implies \theta}$$

Table 4: Mixed rules of **NJ**

1.	$(p \lor q)$	**Prem**
2.	$\neg(p \land r)$	**Prem**
3.	$(q \to s)$	**Prem**
5.	r	**Ass**
7.	p	**Ass**
	$\vdots$	
11.	$(p \land r)$	
12.	$\bot$	2, 11, $\neg\mathbf{E}$
8.	s	12, $\bot\mathbf{E}$
9.	q	**Ass**
10.	s	3, 9, $\to \mathbf{E}$
6.	s	1, 7-8, 9-10, $\lor\mathbf{E}$
4.	$(r \to s)$	5-6, $\to \mathbf{I}$

206

The deduction is then easily completed with $\wedge\mathbf{I}$.

2.5 NK and beyond

This is not the place to give a complete summary of our teaching methodology for natural deduction. We aim for the students to achieve competency in **NJ** within 3 weeks after which they have a test. After that, they are introduced to Classical Natural Deduction (**NK**), with the addition of Double Negation Elimination ($\neg\neg\mathbf{E}$). This is a game-changer. **NJ** still provides the everyday framework within which deductions can be created but now there is a wildcard. Whereas in **NJ** every formula, whether resource or goal, has its own rule, $\neg\neg\mathbf{E}$ can be applied to any goal. Well, even then, our distinction between automatic and choice provides a useful heuristic: $\neg\neg\mathbf{E}$ is never needed for automatic goals.

$\neg\neg\mathbf{E}$ provides the opportunity to introduce a new strategic concept: that of looping. Albert Einstein once defined insanity as doing the same thing and expecting a different result. We use this idea as encouragement: doing a different thing and expecting a different result is not necessarily insane. The classic example of this is the natural deduction proof of excluded middle:

3.	$\neg(p \vee \neg p)$		Ass
7.	p		Ass
	$\vdots$		
8.	$\bot$		
6.	$\neg p$		7-8, $\neg\mathbf{I}$
5.	$(p \vee \neg p)$		6, $\vee\mathbf{I}$
4.	$\bot$		3, 5, $\neg\mathbf{E}$
2.	$\neg\neg(p \vee \neg p)$		3-4, $\neg\mathbf{I}$
1.	$(p \vee \neg p)$		2, $\neg\neg\mathbf{E}$

At this point in the proof you are trying to prove $\bot$, again. At step 4 your goal was also to prove $\bot$. So is this looping insanity? No, because something has changed: you now have line 7, p, as an additional resource, and a solution is only a step away. Contrast this with the following:

3.	$\neg(p \vee \neg p)$		Ass
	$\vdots$		
7.	$(p \vee \neg p)$		
8.	$\bot$		3, 7, $\neg\mathbf{E}$
6.	p		8, $\bot\mathbf{E}$
5.	$(p \vee \neg p)$		6, $\vee\mathbf{I}$
4.	$\bot$		3, 5, $\neg\mathbf{E}$
2.	$\neg\neg(p \vee \neg p)$		3-4, $\neg\mathbf{I}$
1.	$(p \vee \neg p)$		2, $\neg\neg\mathbf{E}$

Here we are trying again to prove $(p \lor \neg p)$. It's the third time this has come up: once on line 1 and again on line 5. But this time we really are looping: for the goal on line 5, our only resource was $\neg(p \lor \neg p)$ on line 3, and again for the goal on line 7, we have only this as a resource. Conclusion: insanity, backtrack.

From **NK** we move on to identity, and consider the complexity of $\forall$**E** and $\exists$**I** in a language with complex terms. This requires a new approach to proof search and introduces various concepts associated with pattern matching. And finally, we move to formal arithmetic, using natural deduction also as a way of teaching mathematical induction. Strategically, this is also interesting because of the need for lemmas in arithmetic. That is another teaching moment on the limits of syntax-guided proof methods.

In summary, natural deduction taught in this way, provides both a training ground for those who want to be able to create their own proofs, but also a wonderful case study of general strategic reasoning, including such topics as the balance between automatic moves and those that require management of choices, the virtues of procrastination, attention to goals and resources, the advantages and disadvantages of too much power ($\neg\neg$**E**), avoiding insanity, and the need for genuine creativity.

3 Natural Deduction Planner

Efficiently creating large numbers of typeset sample deductions can be a daunting prospect. On pen and paper, even a challenging proof can be completed within minutes. However typesetting a proof in software such as LaTeX requires a great deal more effort. With custom packages, structural features such as the scope lines used above can be automated well, but the task of inputting formulas is still cumbersome. Where the hand can draw any symbol at much the speed of any other, typesetting special characters often requires lengthy commands. We began development of a proof assistant software application with the primary goal of overcoming these difficulties, but the result is useful in many more ways than typesetting. We call the result the Natural Deduction Planner (NDP). It generates LaTeX code for use with a custom package.

Our interface essentially replicates the pen and paper proof process, using the same layout and notation of Gentzen's system, as above. Users input sequents using a set of special characters available onscreen. No special formatting (such as prefix notation) is required - a correctly inputted sequent appears as it would on the page. A range of proof systems are available, such as **NJ**, **NK** and Peano Arithmetic. Proofs appear graphically onscreen exactly as they would be typeset.

NDP is similar to the Proof Developer tool created by Daniel Velleman.[3] The interfaces are very alike, but where Proof Developer focusses on informal proof writing, NDP is concerned with formal deductions. Both approaches use ideas of strategy, goals and resources to conduct proofs. Another similar tool is PANDA, developed at the Institut de Recherche en Informatique de Toulouse.[4] PANDA uses a proof tree style, rather than the Fitch-style calculus implemented in NDP.

3.1 Basic Use

The primary use of the Natural Deduction Planner is in the creation of deductions. Computer based deduction systems immediately present a challenge in their particular uses of special symbols. Without a usable interface, inputting formulas can become unnecessarily tedious. For NDP, we found it very important that formulas would appear as they do when handwritten (or as similar as possible). To this end, simplified and easily typed symbol sets (like using &, v and -> for $\wedge$, $\vee$ and $\rightarrow$, respectively) are not the solution. Instead, we implemented a special symbols panel' on the sequent input dialog. This can be seen in Figure 1. Users can input the special symbols by selecting them. A number of intuitive shortcuts (such as `Alt+>` for $\rightarrow$ and `Alt+a` for $\forall$) are also available.

The new proof dialog is designed to reflect the standard style of sequents. Premises are listed in the uppermost textbox, separated by commas. In the lower textbox, the conclusion is supplied. Correct bracket matching must be used, and the system indicates when this is (and is not) the case. Users must input a conclusion in order to begin the deduction, though premises are optional.

In the new proof dialog, users are also able to choose which ruleset to use. In Figure 1, **NJ** has been selected, and so the double negation elimination rule will not be available (though it can be activated during the course of the deduction). The choice of ruleset allows for fine-grained control over how the deduction can proceed. Custom rulesets can be defined, allowing an instructor to, for example, deactivate all quantifier rules. Students can then select the appropriate ruleset for a given exercise, and the deactivated rules will not appear, helping to reduce potential confusion from unknown rules.

Once the sequent has been inputted, the incomplete deduction (premises at the top, with a space before the conclusion) will appear in the main window. At each

---

[3] Proof Developer is a Java web applet built to accompany Velleman's textbook "How to Prove It" [1, 2] `http://www.cs.amherst.edu/~djv/pd/pd.html`

[4] PANDA is a Java application designed for teaching computer science students in logic, developed by Olivier Gasquet, François Schwarzentruber, and Martin Strecker `http://www.irit.fr/panda`

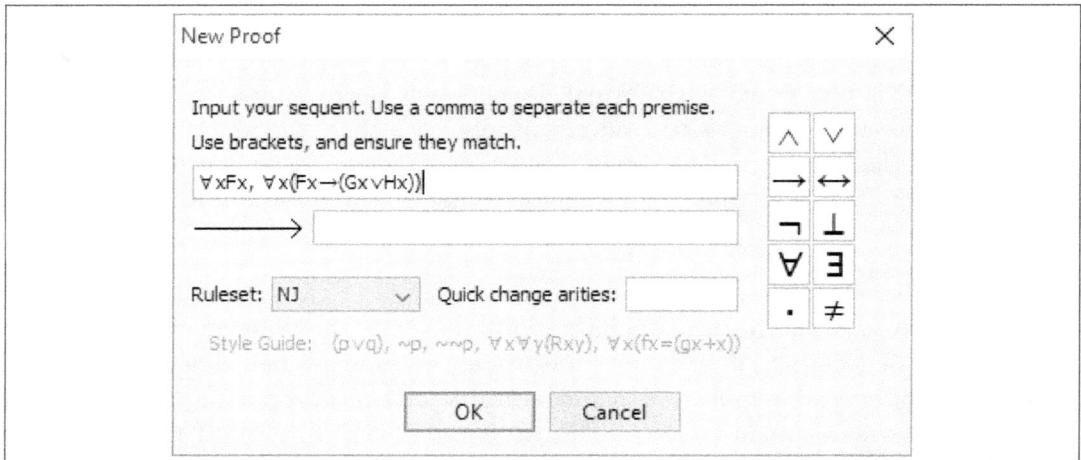

Figure 1: The input dialog. The special symbols panel is on the right and bracket match indication is shown. A conclusion is yet to be inputted.

stage of an NDP deduction, the user can apply any valid rules, and their outcome is immediately displayed. To apply a rule, a *current goal* must first be selected, which can be any unjustified line (indicated by a missing justification). This can be achieved by simply clicking on the line in question. Upon selecting a current goal, that line's introduction rules (if any) appear as button(s) next to it, and the line is highlighted. To apply one of these rules, a user can simply select that button. This takes care of introduction rules. For elimination rules, the user must also select a relevant *current resource*. Upon selection, the resource is highlighted and its elimination rules appear alongside the current goal. Again, to use a rule it is simply selected.

NDP checks the scope of lines available to the current goal. If a user selects a resource out of scope of the current goal, its elimination rules appear "greyed out", and cannot be applied. In this way, NDP ensures the user is following rules correctly. This behaviour is useful in teaching students the details of scope. An extension of this is shown in Figure 2. Here lines out of scope of the goal are greyed out, to explicitly show that they cannot be used. This feature can be activated from the Options menu. It does not prevent users from selecting out-of-scope resources, but does make it very clear how scope works.

Lines are numbered in the order of creation, again reinforcing the way the deduction is constructed. This is also useful when reaching a dead-end in the proof - the user can see exactly how they reached this point, and what will happen when they retrace their steps. Highlighting is used to indicate the current goal and resource

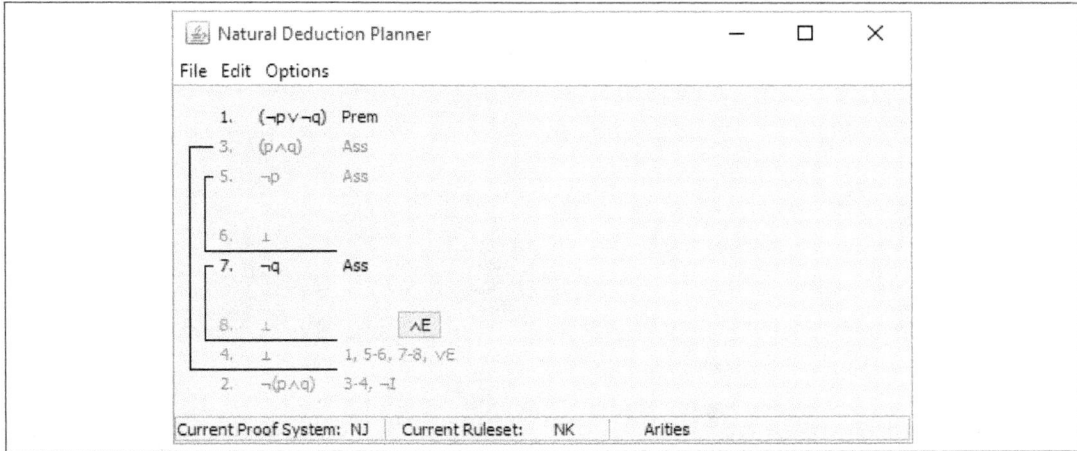

Figure 2: An incomplete proof showing current goal (green), current resource (red) and possible rules ($\wedge$**E**). All **NK** rules are available, but currently only **NJ** rules have been used.

and also serves to indicate any lines out of scope of the goal. Once all lines have been justified the proof is complete.

The requirement that users select a current goal before choosing a resource is fundamental to the operation of NDP. Even if the first step in the proof involves an elimination rule, the user must select a current goal. These explicit realisations of goal and resource help to reinforce the use of strategies in deductions. Rather than beginning with premises and working downwards, the user is encouraged to begin at the bottom of the deduction (the goal) and efficiently choose those resources which are needed. By breaking from a strictly linear approach, selecting goals encourages users to consider which available rules are automatic, and which require choice.

3.2   Types of Rules

The rules of natural deduction can be broadly assigned to one of three categories: those which are completely automatic, being applied the same way in every instance (for example, $\wedge$**I**); those requiring some level of choice, as between two options (e.g. $\vee$**I**); and those which require further input, such as a choice of term (e.g. $\exists$**I**). Of course, further distinctions can be made between these categories, and the boundaries between them are fairly blurred ($\rightarrow$ **E**, for example, would be in the first or second categories, depending on the case).

Knowing which rules fall into which category is very important for developing

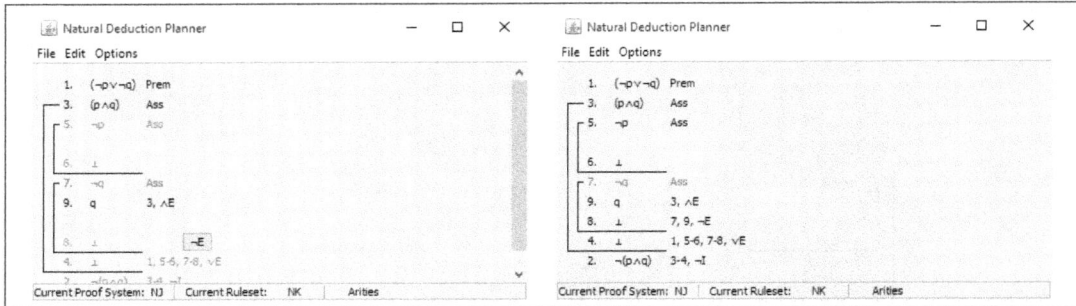

Figure 3: ¬**E** before application (left) and after (right). The user has clicked the button ¬**E** and done nothing else.

the skill of creating deductions effectively. The automatic rules should be applied quickly, and choice rules with caution. While this can be explained to students, often experience is the best way to reinforce just why we should choose one rule over another. As a direct consequence of the implementation of each rule in NDP, it becomes very obvious through use which rules are automatic and which are not. As such, practice with NDP can help students learn the best orders of rule application.

How does the behaviour of the types of rule differ in NDP? All rules are applied by selecting the appropriate button next to the current goal. They differ in what happens next.

For the automatic rules, the proof is immediately updated. Since there is no further information needed to apply the rule, no further interaction with the user is required. Figure 3 shows the application of the automatic ¬**E** rule. Negation elimination behaves differently depending on the context. Here, the goal is a contradiction ($\perp$) and the negand (p) can be found within scope. So all that needs to be given is a justification for $\perp$. In fact, this rule is *always* automatic, no matter the context. If the current goal is not $\perp$ then $\perp$ will be introduced, justified and the current goal justified by explosion ($\perp$**E**). If the negand does not already appear in the proof, it will be added as the new current goal above $\perp$.

A generally automatic rule which blurs the distinction between automation and choice is ∧**E**, which is applied to a current resource. If the current goal is one of the conjuncts, then it is automatically justified and nothing further happens. Otherwise, the user is presented with a (fairly trivial) choice: would you like to extract the first conjunct, the second conjunct or both? This is not a difficult choice - if in doubt we can choose both and get, at worst, a proof of one line longer. But this choice, presented to the user, reinforces the bigger picture. We're using strategy to move through the deduction, and it may be possible to proceed with only one of the

212

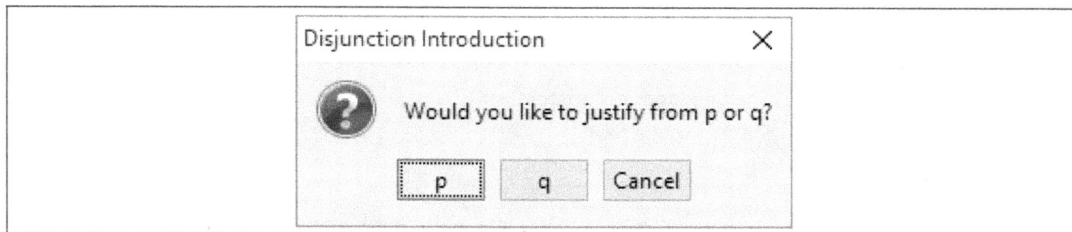

Figure 4: The disjunction introduction dialog.

conjuncts. The choice reminds the user to think ahead.

A classic example of a choice rule is $\vee\mathbf{I}$, disjunction introduction. Here, a choice must be made as to which disjunct to prove from (at least in **NJ** - under **NK** we could also double negate). In NDP, the user must choose how to proceed. This is implemented through a dialog as shown in Figure 4. This dialog breaks the flow of the proof. Where users click through automatic rules with little thought, disjunction introduction requires more input. This explicit demonstration of the location of choice helps to show when it is required, and why.

An example of the third category of rules, those which require explicit further input from the user, is existential introduction $\exists\mathbf{I}$. To justify $\exists x Fx$ the user must choose a term to justify from (assuming the currently selected resource does not match the pattern Fx). The user is again presented with a dialog, but now must manually input a term to use, instead of simply choosing another button. Figure 5 shows the dialog presented in this case. The term input contains free syntax, and can be as long as required, allowing for the input of functional terms (e.g. $ffgfaffb$). Say the user inputs a, yielding Fa. NDP will then check to see if Fa already appears in scope. If it does, the current goal will be justified. Otherwise, Fa will be added as a new goal. A similar procedure applies for $\forall\mathbf{E}$. In this case, however, the choice of term is fairly harmless - as with $\wedge\mathbf{E}$ we can always re-use a universal resource. The distinction between the choices of $\exists\mathbf{I}$ and $\forall\mathbf{E}$ become clear in NDP, since once $\exists\mathbf{I}$ is applied that line cannot be selected again, whereas the $\forall$ line can.

3.3 Level of Automation

A standard feature of much proof assistant software is automation of the proof process. NDP, by contrast, has been built with very little automation in mind. The automation implemented is at the level of rules, rather than proofs. That is, we have tried to automate the application of each rule as much as possible, without automating the proof process itself. The reason for this is that the focus is on replicating the pen-and-paper process.

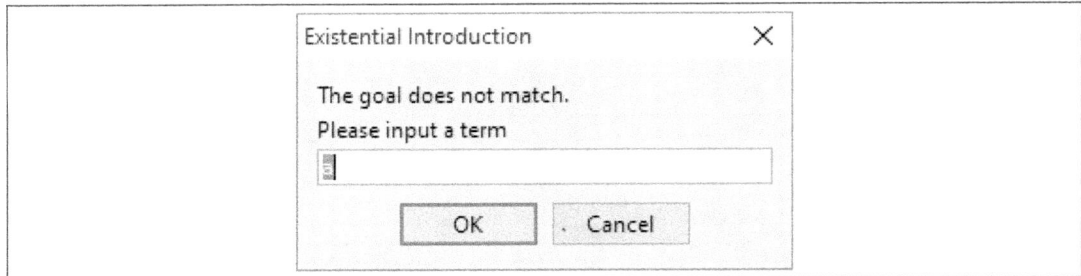

Figure 5: The existential introduction dialog. The user has inputted a.

By automating the writing of each proof line, users are able to move through deductions faster, focussing more on the strategy involved. Speed and the ability to easily "undo" mistakes also removes hurdles from the bulk trial and error method of learning strategy. A student unsure of the next step in a complicated pen and paper proof may be overly wary - a wrong move would result in writing out the whole proof again. In NDP, however, she can chose a rule in the knowledge that the current proof state can easily be retrieved. Similarly, students sometimes find rules like disjunction elimination, which requires creating four new lines and two scopes, to be intimidating and tiresome. Yet disjunction elimination is an automatic rule and should be applied as quickly as possible. In NDP disjunction elimination is achieved with two clicks, a less daunting task.

NDP applies rules, but the user must directly control it. Of course in some cases, the automation can circumvent attempts at learning. For example, the universal introduction rule always generates a new constant, that appears nowhere else in the proof. By doing so, the requirements for terms with that rule are always met. However, this takes some control away from the user. The $\forall$I rule does not always require a *completely* new term - if one appears out of scope it can be reused. NDP includes an option to disable automatic parameters. If a user chooses to do this, they will be asked for a term when performing $\forall$I and $\exists$**E**. If this term is illegal (violates the requirements for terms with those rules), the user is notified, and the offending justification marked.

3.4 Further Features

A *Rule Palette* allows individual rules to be (de)activated independently of the proof system chosen. The rule palette's layout shows the symmetry of Gentzen's rules, and gives some indication as to how the rules fit into different logical systems. By only activating certain rules, students can complete exercises in subsets of a system before being introduced to it fully, and see where certain rules are needed. For

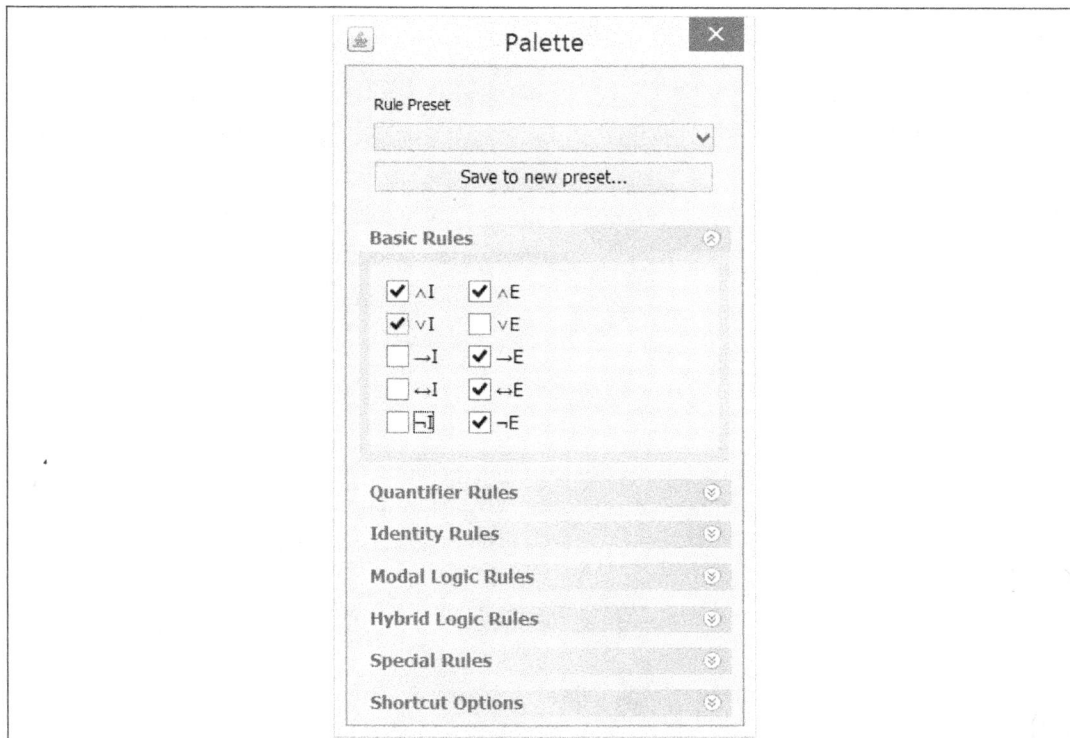

Figure 6: The Rule Palette. Only simple deduction rules are available; rules involving assumptions have been deactivated.

example, the rule palette can be used to demonstrate the importance of double negation elimination in **NK**, by attempting a proof of $(p \lor \neg p)$ without double negation elimination to see how far it goes. Once we get stuck, we turn on double negation elimination to finish the proof. Users can try out their own systems too, to see how different rules interact with each other.

Upon finishing a proof, it can be saved as either an editable proof or a demonstration proof. A demonstration proof has interaction disabled, providing a means to follow through an already complete proof. This is essentially the step by step deduction given above but in electronic form. Editable proofs behave similarly, but allow a user to take over the proof at any point. No work further than completing the deduction is required to generate these. Proofs can also be exported to unicode format and as an image. Complete proofs can also easily be animated in `.gif` format, for use in slides or online. A primary feature of NDP is its ability to export proofs to LaTeX code. This interacts with a LaTeX package (based on Ti*k*Z) which

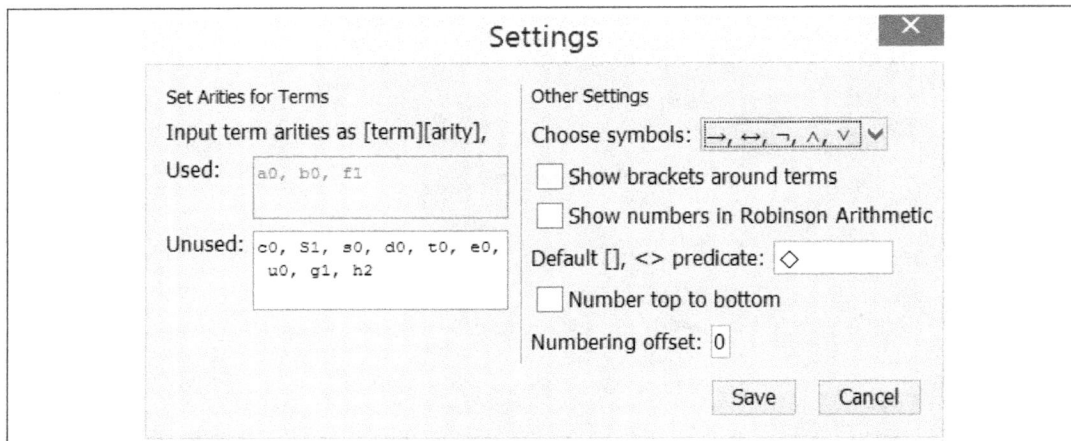

Figure 7: The settings dialog. The constants a and b have been used in the proof, as has the unary function f. It is still possible to change the arities of c, S, s, d, t, e, u, g and h. "Show numbers in Robinson Arithmetic" causes terms like $SSS0$ to be displayed as 3.

generates nicely typeset deductions. The task of producing exercises and their solutions involves little more than completing deductions on NDP - no fiddling about with alignment or trying to recall commands required.

In order to increase flexibility in using NDP, a number of settings are available. Two symbol sets ($\neg, \wedge, \vee, \rightarrow, \leftrightarrow$ vs. $\sim, \&, \vee, \supset, \equiv$) can be chosen from. The line numbering can be tweaked in two ways. Standard top-to-bottom numbering is possible, and an offset can be applied, so that sub-proofs can be replaced without repeating a proof in full. Figure 7 shows the settings dialog.

Though originally intended to cover only propositional and predicate logic and Peano Arithmetic, we have begun extending NDP to cover other logics, and to consider new features. We've implemented a system of modal logic and hybrid logic using a labelled deduction method. These rules are available in the standard rule palette but are not thoroughly tested. A very rudimentary second order logic is also available, easily implemented due to Java seeing no distinction between predicate and variable symbols when making substitutions. In an attempt to automate the proof process, a *Magic Mode* is provided. This applies any rules which require no extra input for up to 10 iterations. In exceptional circumstances Magic Mode can complete proofs, but in general will only move forward one or two steps. Finally, a method to include custom axioms has been implemented.

3.5   Use in Teaching

We have used NDP as part of a course teaching natural deduction strategies. All the deduction exercises for the course were generated by the software, and it was also made available for students to download. Many students did so, and used NDP to complete exercises and study for tests. We released exercise solutions both as text documents and editable proof files. A novel use NDP was put to was in catching up on missed lectures. Since NDP applies each rule correctly, by studying what happens students could learn the rule themselves. While the motivation and strategy discussed in lectures was absent here, the correct manipulation of the formula was learned. NDP's automated rule application had some downsides though. Some students found overuse of NDP resulted in over reliance - you don't have to remember how to set out implication introduction if the software does it for you. Since tests were by pen and paper, this proved problematic. The best combination seemed to be use of pen and paper to practise rules, and NDP to practise strategy.

In the context of tutorials, NDP allowed for greater flexibility in presentation. Again due to "undo" it was easier to recover from bad choices, encouraging student participation. Also, a source of potentially confusing transcription errors - the tutor's handwriting - was removed. In one on one situations, NDP allowed for a greater flow of conversation. Discussed strategies for stuck proofs could be implemented quickly and results considered in much less time than would be required to write 10 lines of formulas by hand.

3.6   Implementation

NDP was implemented in Java using the **Swing** and **SwingX** graphical user interface libraries. The code was written to be extendable and with a goal of modularity, to allow different interfaces to interact with the same backend.

Formulas are held as strings in a TeX macro format, using prefix notation for easier argument parsing. This also simplifies the process of exporting proofs to LaTeX code. For example, the formula $(Fa \land \forall x(Fx \to Gx))$ would be stored as `\con{Fa}{\qa{x}{\imp{Fx}{Gx}}}`. Each line of a proof is an **NDLine** object, which contains information such as the formula, the line number, the type of line and the justification. The **NDLine** class also has methods returning each argument of a formula.

The **ProofMethods** class forms the core of the program. This holds the current proof state as an array of **NDLines**. The application of a rule results in a relevant modification of the proof array and any lines within it. For example, suppose $\land \mathbf{E}$ is applied, as possible in Figure 2. Line 3 is the resource, and **ProofMethods** first obtains line 3's arguments (p and q) from its **NDLine**. Neither p nor q matches the current goal (line 8, $\perp$), and so the user will be asked whether they wish to extract both p and q, or just one. Suppose the user selects to only extract q (this seems a good move, since we'll be able to get a contradiction with $\neg q$). The current proof state will be extended with a new **NDLine** containing q, justified with 3, $\wedge\mathbf{E}$.

Rules themselves are methods within the **ProofMethods** class, and the system can be extended by adding new methods to give new rules. In practice, this means that additional rules (such as Disjunctive Syllogism) or extensions to the system can be added fairly easily. **ProofMethods** is designed to be as self-contained as possible; methods for printing the proof array to the command line mean it could be used without a graphical interface. In fact, this is how early development proceeded. Unfortunately, **ProofMethods** is not entirely standalone. Specifically, when interaction from the user is required (such as choosing a term for $\forall E$), **ProofMethods** must fall back on **Swing** libraries for graphical input.

On top of **ProofMethods** sits the **ProofPanel** class, a modified **JPanel** which provides user interaction with **ProofMethods**. **ProofPanel** interprets the proof array and arranges the deduction onscreen. The function of the rule palette is implemented entirely within the **ProofPanel**. If conjunction introduction ($\wedge\mathbf{I}$) is turned off then the option to apply that rule becomes unavailable on the **ProofPanel**. That is, even with $\wedge\mathbf{I}$ "disabled" **ProofMethods** is still able to apply that rule - there is just no way for the command to do so to reach it. The rule palette makes extensive use of the **SwingX** library.

A modified **JFrame** constitutes the main window of the Proof Assistant and controls tasks such as New Proof, Save, Open and Export. Proofs are saved in plain text files which contain complete undo histories and settings profiles. There is no difference between **.ndp** (editable) and **.ndu** (demonstration) files — they are read in differently but their contents are identical.

NDP is available on SourceForge at `http://sourceforge.net/p/proofassistant/`.

References

[1] D. Velleman, *How to Prove it: A Structured Approach*. Cambridge University, 1994.

[2] D. Velleman, *How to Prove It: A Structured Approach*. Cambridge University Press, 2006.

[3] A. Cupillari, *The Nuts and Bolts of Proofs*. Elsevier Academic Press, 2005.

[4] D. Solow, *How to read and do proofs: an introduction to mathematical thought processes*. Wiley, 2002.

[5] I. Copi, C. Cohen, and K. McMahon, *Introduction to Logic: Pearson New International Edition*. Pearson Education, Limited, 2013.

Received 11 October 2016